Comparing and Measuring

TEACHER'S GUIDE

SCIENCE AND TECHNOLOGY FOR CHILDREN™

NATIONAL SCIENCE RESOURCES CENTER
Smithsonian Institution • National Academy of Sciences
Arts and Industries Building, Room 1201
Washington, DC 20560

NSRC

The National Science Resources Center is operated by the Smithsonian Institution and the National Academy of Sciences to improve the teaching of science in the nation's schools. The NSRC collects and disseminates information about exemplary teaching resources, develops and disseminates curriculum materials, and sponsors outreach activities, specifically in the areas of leadership development and technical assistance, to help school districts develop and sustain hands-on science programs.

STC Project Supporters

National Science Foundation
Smithsonian Institution
U.S. Department of Defense
U.S. Department of Education
John D. and Catherine T. MacArthur Foundation
The Dow Chemical Company Foundation
E. I. du Pont de Nemours & Company
Amoco Foundation, Inc.
Hewlett-Packard Company
Smithsonian Institution Educational Outreach Fund
Smithsonian Women's Committee

This project was supported, in part, by the
National Science Foundation
Opinions expressed are those of the authors and not necessarily those of the Foundation

02 01 00 99   10 9 8 7 6 5 4 3

ISBN 0-89278-609-4

Published by Carolina Biological Supply Company, 2700 York Road, Burlington, NC 27215.
Call toll free 1-800-334-5551.

This material is based upon work supported by the National Science Foundation under Grant No. ESI-9252947. Any opinions, findings, and conclusions or recommendations expressed in this material are those of the author(s) and do not necessarily reflect the views of the National Science Foundation.

CB787429904

Printed on recycled paper.

Foreword

Since 1988, the National Science Resources Center (NSRC) has been developing Science and Technology for Children (STC), an innovative hands-on science program for children in grades one through six. The 24 units of the STC program, four for each grade level, are designed to provide all students with stimulating experiences in the life, earth, and physical sciences and technology while simultaneously developing their critical-thinking and problem-solving skills.

Sequence of STC Units

Grade	Life, Earth, and Physical Sciences and Technology			
1	*Organisms*	*Weather*	*Solids and Liquids*	*Comparing and Measuring*
2	*The Life Cycle of Butterflies*	*Soils*	*Changes*	*Balancing and Weighing*
3	*Plant Growth and Development*	*Rocks and Minerals*	*Chemical Tests*	*Sound*
4	*Animal Studies*	*Land and Water*	*Electric Circuits*	*Motion and Design*
5	*Microworlds*	*Ecosystems*	*Food Chemistry*	*Floating and Sinking*
6	*Experiments with Plants*	*Measuring Time*	*Magnets and Motors*	*The Technology of Paper*

The STC units provide children with the opportunity to learn age-appropriate concepts and skills and to acquire scientific attitudes and habits of mind. In the primary grades, children begin their study of science by observing, measuring, and identifying properties. Then they move on through a progression of experiences that culminate in grade six with the design of controlled experiments.

Sequence of Development of Scientific Reasoning Skills

Scientific Reasoning Skills	Grades					
	1	2	3	4	5	6
Observing, Measuring, and Identifying Properties	◆	◆	◆	◆	◆	◆
Seeking Evidence Recognizing Patterns and Cycles		◆	◆	◆	◆	◆
Identifying Cause and Effect Extending the Senses				◆	◆	◆
Designing and Conducting Controlled Experiments						◆

The "Focus-Explore-Reflect-Apply" learning cycle incorporated into the STC units is based on research findings about children's learning. These findings indicate that knowledge is actively constructed by each learner and that children learn science best in a hands-on experimental environment where they can make their own discoveries. The steps of the learning cycle are as follows:

- Focus: Explore and clarify the ideas that children already have about the topic.
- Explore: Enable children to engage in hands-on explorations of the objects, organisms, and science phenomena to be investigated.
- Reflect: Encourage children to discuss their observations and to reconcile their ideas.
- Apply: Help children discuss and apply their new ideas in new situations.

The learning cycle in STC units gives students opportunities to develop increased understanding of important scientific concepts and to develop positive attitudes toward science.

The STC units provide teachers with a variety of strategies with which to assess student learning. The STC units also offer teachers opportunities to link the teaching of science with the development of skills in mathematics, language arts, and social studies. In addition, the STC units encourage the use of cooperative learning to help students develop the valuable skill of working together.

In the extensive research and development process used with all STC units, scientists and educators, including experienced elementary school teachers, act as consultants to teacher-developers, who research, trial teach, and write the units. The process begins with the developer researching the unit's content and pedagogy. Then, before writing the unit, the developer trial teaches lessons in public school classrooms in the metropolitan Washington, D.C., area. Once a unit is written, the NSRC evaluates its effectiveness with children by field-testing it nationally in ethnically diverse urban, rural, and suburban public schools. At the field-testing stage, the assessment sections in each unit are also evaluated by the Program Evaluation and Research Group of Lesley College, located in Cambridge, Mass. The final editions of the units reflect the incorporation of teacher and student field-test feedback and of comments on accuracy and soundness from the leading scientists and science educators who serve on the STC Advisory Panel.

The STC project would not have been possible without the generous support of numerous federal agencies, private foundations, and corporations. Supporters include the National Science Foundation, the Smithsonian Institution, the U.S. Department of Defense, the U.S. Department of Education, the John D. and Catherine T. MacArthur Foundation, the Dow Chemical Company Foundation, the Amoco Foundation, Inc., E. I. du Pont de Nemours & Company, the Hewlett-Packard Company, the Smithsonian Institution Educational Outreach Fund, and the Smithsonian Women's Committee.

Acknowledgments

Comparing and Measuring was researched and developed by Christopher Lyon, edited by Lynn Miller, and illustrated by Lois Sloan. Other NSRC staff who contributed to the development of this unit include Charles N. Hardy, deputy director for information dissemination, materials development, and publications (1995–96); Sally Goetz Shuler, deputy director for development, external relations, and outreach; Joyce Lowry Weiskopf, STC project director (1992–95); Katherine Stiles, STC Research Associate (1990–95); Dean Trackman, publications director; and Heidi M. Kupke, publications technology specialist. The unit was evaluated by Sabra Lee, senior research associate, Program Evaluation and Research Group, Lesley College, Boston, MA. *Comparing and Measuring* was trial taught in Jane Frydenlund's first-grade classroom at Annandale Terrace Elementary School in Annandale, Virginia.

The technical review of *Comparing and Measuring* was conducted by Duane A. Cooper, Assistant Professor, Center for Mathematics Education, University of Maryland, College Park, MD.

The NSRC would like to thank the following individuals and school systems for their assistance with the national field-testing of the unit:

Brazosport Independent School District, Brazosport, TX
Coordinator: Donna Jablecki, Science and Health Supervisor
Margaret Cast, Teacher, O. A. Fleming Elementary School
Kay Eitel, Teacher, Gladys Polk Elementary School
Frances Parks, Teacher, S. F. Austin Elementary School
Karen Suggs, Teacher, O. A. Fleming Elementary School

The Einstein Project, Green Bay, WI
Coordinator: Cecilia Turriff
Green Bay Area Public School District
Julie Bogden, Teacher, Howe Elementary School
Ashwaubenon School District, Ashwaubenon, WI
Wendy Ohlhues, Teacher, Pioneer Elementary School
Oneida, WI
Darlene Schoen, Teacher, Oneida Nation Elementary School

North Penn School District, Lansdale, PA
Coordinator: Donna Brown, Science Supervisor
David Decker, Math Supervisor
Caroline Crew, Teacher, Gwynedd Square Elementary School
Susan Delp, Teacher, Hatfield Elementary School
Nancy Garis, Teacher, Montgomery Elementary School

Spokane School District No. 81, Spokane, WA
Coordinator: Scott Stowell, Science Supervisor
Diane Campbell, Teacher, Jefferson Elementary School
Barbara Edwards, Teacher, Woodbridge Elementary School
Diane Figueroa, Teacher, Holmes Elementary School

The NSRC also would like to thank the following individuals for their contributions to the unit:

Jane Frydenlund, Teacher, Annandale Terrace Elementary School, Annandale, VA
Mary Ellen McCaffrey and David Burgevin, Photographic Production Control, Office of Printing and Photographic Services, Smithsonian Institution, Washington, DC
Philip Morrison, Professor of Physics Emeritus, Massachusetts Institute of Technology, Cambridge, MA
Phylis Morrison, Educational Consultant, Cambridge, MA
Trudie Ogilvie, Educational Assistant, Annandale Terrace Elementary School, Annandale, VA
Dane Penland, Chief, Imaging and Technology Services Branch, Office of Printing and Photographic Services, Smithsonian Institution, Washington, DC
Richard Strauss, Photographer, Office of Printing and Photographic Services, Smithsonian Institution, Washington, DC
Jeff Tinsley, Chief, Special Assignments/Photography Branch, Office of Printing and Photographic Services, Smithsonian Institution, Washington, DC

Douglas Lapp
Executive Director
National Science Resources Center

STC Advisory Panel

Peter P. Afflerbach, Professor, National Reading Research Center, University of Maryland, College Park, MD

David Babcock, Director, Board of Cooperative Educational Services, Second Supervisory District, Monroe-Orleans Counties, Spencerport, NY

Judi Backman, Math/Science Coordinator, Highline Public Schools, Seattle, WA

Albert V. Baez, President, Vivamos Mejor/USA, Greenbrae, CA

Andrew R. Barron, Professor of Chemistry and Material Science, Department of Chemistry, Rice University, Houston, TX

DeAnna Banks Beane, Project Director, YouthALIVE, Association of Science-Technology Centers, Washington, DC

Audrey Champagne, Professor of Chemistry and Education, and Chair, Educational Theory and Practice, School of Education, State University of New York at Albany, Albany, NY

Sally Crissman, Faculty Member, Lower School, Shady Hill School, Cambridge, MA

Gregory Crosby, National Program Leader, U.S. Department of Agriculture Extension Service/4-H, Washington, DC

JoAnn E. DeMaria, Teacher, Hutchison Elementary School, Herndon, VA

Hubert M. Dyasi, Director, The Workshop Center, City College School of Education (The City University of New York), New York, NY

Timothy H. Goldsmith, Professor of Biology, Yale University, New Haven, CT

Patricia Jacobberger Jellison, Geologist, National Air and Space Museum, Smithsonian Institution, Washington, DC

Patricia Lauber, Author, Weston, CT

John Layman, Director, Science Teaching Center, and Professor, Departments of Education and Physics, University of Maryland, College Park, MD

Sally Love, Museum Specialist, National Museum of Natural History, Smithsonian Institution, Washington, DC

Phyllis R. Marcuccio, Associate Executive Director for Publications, National Science Teachers Association, Arlington, VA

Lynn Margulis, Distinguished University Professor, Department of Botany, University of Massachusetts, Amherst, MA

Margo A. Mastropieri, Co-Director, Mainstreaming Handicapped Students in Science Project, Purdue University, West Lafayette, IN

Richard McQueen, Teacher/Learning Manager, Alpha High School, Gresham, OR

Alan Mehler, Professor, Department of Biochemistry and Molecular Science, College of Medicine, Howard University, Washington, DC

Philip Morrison, Professor of Physics Emeritus, Massachusetts Institute of Technology, Cambridge, MA

Phylis Morrison, Educational Consultant, Cambridge, MA

Fran Nankin, Editor, *SuperScience Red*, Scholastic, New York, NY

Harold Pratt, Senior Program Officer, Development of National Science Education Standards Project, National Academy of Sciences, Washington, DC

Wayne E. Ransom, Program Director, Informal Science Education Program, National Science Foundation, Washington, DC

David Reuther, Editor-in-Chief and Senior Vice President, William Morrow Books, New York, NY

Robert Ridky, Professor, Department of Geology, University of Maryland, College Park, MD

F. James Rutherford, Chief Education Officer and Director, Project 2061, American Association for the Advancement of Science, Washington, DC

David Savage, Assistant Principal, Rolling Terrace Elementary School, Montgomery County Public Schools, Rockville, MD

Thomas E. Scruggs, Co-Director, Mainstreaming Handicapped Students in Science Project, Purdue University, West Lafayette, IN

Larry Small, Science/Health Coordinator, Schaumburg School District 54, Schaumburg, IL

Michelle Smith, Publications Director, Office of Elementary and Secondary Education, Smithsonian Institution, Washington, DC

Susan Sprague, Director of Science and Social Studies, Mesa Public Schools, Mesa, AZ

Arthur Sussman, Director, Far West Regional Consortium for Science and Mathematics, Far West Laboratory, San Francisco, CA

Emma Walton, Program Director, Presidential Awards, National Science Foundation, Washington, DC, and Past President, National Science Supervisors Association

Paul H. Williams, Director, Center for Biology Education, and Professor, Department of Plant Pathology, University of Wisconsin, Madison, WI

car
piano
conference table

Contents

Goals for *Comparing and Measuring*

In this unit, students' observations and activities expand their awareness of comparing and measuring. From their experiences, they are introduced to the following concepts, skills, and attitudes.

Concepts

- Comparing involves observing similarities and differences.
- One way to make comparisons is by matching.
- Using beginning and ending points and placing units end to end are important factors when measuring.
- Nonstandard units of measure produce varying results.
- Standard units of measure produce more consistent results than nonstandard units and make it possible to share information.
- Different units and tools can be used to measure objects.
- Long tools make it easier to measure long objects.
- A common starting line is required to make fair comparisons.

Skills

- Observing similarities and differences among objects.
- Describing similarities and differences among objects.
- Placing objects in serial order on the basis of height or length.
- Communicating observations, ideas, and questions through discussion, drawing, and writing.
- Organizing information on representational graphs and charts.
- Making predictions about the relative lengths and sizes of objects.
- Using standard and nonstandard units of measure.
- Using groups of tens to quantify large numbers of units.
- Measuring using beginning and ending points.
- Interpreting results of measurements.

Attitudes

- Developing an awareness of self and others by comparing height, length of arms and legs, and body cutouts.
- Developing an appreciation of the usefulness of measuring in our daily lives.
- Becoming comfortable using a variety of measuring tools and units of measure.
- Recognizing the importance of developing strategies for counting large numbers.
- Appreciating the importance of organizing information on graphs and charts.

Unit Overview and Materials List

Children naturally make comparisons. At one time or another, many children have stood back to back with a friend to find out who is taller. They have placed their feet next to each other to find out whose foot is longer. They have gone shopping and "matched" their bodies to different sized clothing to find clothes that fit. All these experiences involve comparing, which lays the foundation for matching and, subsequently, learning to measure.

Comparing and measuring are important science skills. When scientists do experiments, they often need to measure; that is, using numbers and standard units of measure, they describe such properties as length, volume, weight, and temperature. Similarly, when students do classroom experiments, they will need to have developed skills in measuring these properties. Even outside the classroom, students have frequent opportunities to measure and compare; for example, when they assemble toy models, fit pieces into jigsaw puzzles, or chart their heights. In fact, comparing and measuring are key ways children make sense of their lives.

Comparing and Measuring, a 16-lesson unit designed for first-graders, gives students a variety of experiences in comparing, matching, and measuring. Throughout the unit, students will observe similarities and differences among objects and match and measure lengths, heights, and distances.

Lesson 1 begins with a class brainstorming session in which students discuss what it means to compare. They begin to recognize that when they identify similarities and differences they are comparing. As a pre-unit assessment lesson, it provides you with a sense of the kinds of comparisons students are making at present and the methods they use to make them.

In the next three lessons, students continue to explore making comparisons. In Lesson 2, they make life-size cutouts of their bodies and use the cutouts to compare their heights. This experience prepares students for matching their heights with adding machine tape in Lesson 3. Students begin to see that when matching the tape to an object, they need to determine both a beginning and an ending point. In Lesson 4, students match the lengths of their arms and legs and record their results on a representational bar graph. When comparing the information on the arm and leg graphs with the information on the height graph from Lesson 3, students begin to recognize the importance of a common starting line.

In the first four lessons, students compare and match lengths either their own size or smaller. In Lesson 5, students are challenged to match larger objects. As they discuss which object might be the largest, students are introduced to the idea of making predictions. In Lesson 6, students begin to discover that matching distance is another facet of matching length. They flip toy Flippers™ and use adding machine tape to represent the distance the Flippers™ have traveled.

In Lesson 7, students make the transition from matching to measuring length by quantifying nonstandard units of measure—in this case, their own feet. Through their measuring activities and a reading selection, students are introduced to the idea that using nonstandard units of measure produces varied results. In Lesson 8, students continue to get varied results as they use different sets of standard objects to measure the same lengths. For example, students who use pencils to measure an object achieve different results from those who use wooden spools to measure it. In addition, students now begin to think about why it is helpful to label their results with the name of the unit they have used to measure. These experiences lay the foundation for Lesson 9, in which the entire class uses the same standard unit to measure. In addition to discovering less variation in their results, students begin to see that using a common standard unit produces results that can be expressed using a common language.

In Lesson 10, students are introduced to a measuring unit that they will use in the next six lessons—Unifix Cubes™. As students connect and stack the cubes, they determine that the cubes are a more versatile tool and that they can be used to measure vertically as well as horizontally. In Lessons 11 and 12, students use larger numbers of cubes to measure larger objects, and gain experience in counting by groups of 10.

In Lesson 13, students make a measuring tool that represents 10 Unifix Cubes™. Students discuss how this tool—a paper measuring strip—eliminates the need for actual Unifix Cubes™. In Lesson 14, students face the challenge of measuring objects that are longer than their measuring strips. Lesson 15 builds on the idea that using an appropriate measuring tool makes it easy to measure long objects. Students create a 100-unit measuring tape and listen to a reading selection to reinforce this concept. In addition, students discover that the measuring tape enables them to measure circumference as well as the length or height of an object.

In the final lesson, students use their measuring tapes to measure how far they can make the Flippers™ travel. Students expand their awareness that measuring is an extension of matching when they compare the methods and results recorded in Lesson 6 with those in this lesson. Lesson 16 also serves as an embedded assessment of the comparing, matching, and measuring skills students have learned during the unit.

Following Lesson 16 is a post-unit assessment that is matched to the pre-unit assessment in Lesson 1. Additional assessments offer students further challenges in comparing, matching, and measuring that will enable you to evaluate students' progress.

After completing this unit, your students will have had the opportunity to develop a basic understanding of key measuring concepts, including using beginning and ending points, using a common starting line, and recognizing the importance of using standard units when measuring. This understanding prepares students for the introduction of formal measuring and the use of units such as centimeters and inches.

Materials List

Below is the list of materials needed for the *Comparing and Measuring* unit. Please note that the metric and English equivalent measurements in this unit are approximate.

1 *Comparing and Measuring* Teacher's Guide
*30 optional Student Notebooks (*My Comparing and Measuring Book*)
15 rolls of pink adding machine tape, 6 cm × 50 m (2½ in × 165 ft)
15 rolls of yellow adding machine tape, 6 cm × 50 m (2½ in × 165 ft)
1 red marker
30 Flippers™ (15 red, 15 blue)
30 resealable plastic bags, 23 × 30 cm (9 × 12 in)
15 red crayons
15 blue crayons
1 roll of masking tape
250 wood coffee stirrers
100 unsharpened pencils
100 plastic spoons
100 toothpicks
100 small wood spools, 4 cm (1½ in)
1 pad of 100 Post-it™ notes, 8 × 13 cm (3 × 5 in)
1,500 Unifix Cubes™ (750 red, 750 blue)
** Assorted markers
** Crayons
** Scissors
** Newsprint for all charts
** Glue
**30 large sheets of paper to trace body outlines (see note)

***Note:** The optional Student Notebooks are available from Carolina Biological Supply Company (1-800-334-5551).

****Note:** These items are not included in the kit. They are available in most schools or can be brought from home. Teachers have used bulletin board or craft paper for the paper to trace body outlines.

Teaching *Comparing and Measuring*

The following information on unit structure, teaching strategies, materials, and assessment will help you give students the guidance they need to make the most of their hands-on experiences with this unit.

Unit Structure

How Lessons Are Organized in the Teacher's Guide: Each lesson in the *Comparing and Measuring* Teacher's Guide provides you with a brief overview, lesson objectives, key background information, a materials list, advance preparation instructions, step-by-step procedures, and helpful management tips. Many of the lessons include recommended guidelines for assessment. Lessons also frequently indicate opportunities for curriculum integration. Look for the following icons that highlight extension ideas:

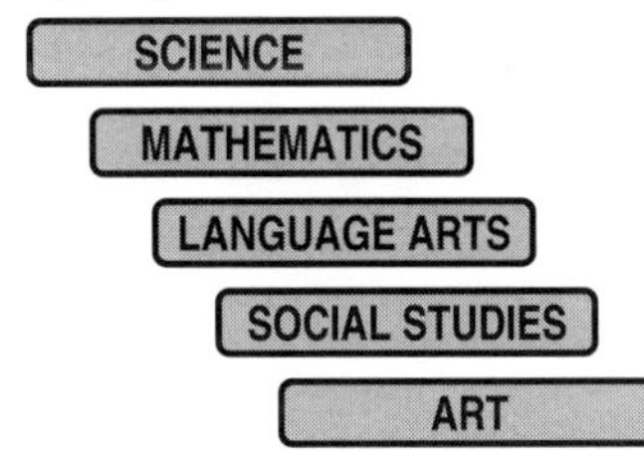

Please note that all record sheets, blackline masters, student instructions, and reading selections may be copied and used in conjunction with the teaching of this unit.

The *Comparing and Measuring* Student Notebook: An optional consumable notebook for students, *My Comparing and Measuring Book,* has been published for this unit. The notebook is an individual, bound copy of all the record sheets contained in the Teacher's Guide. If your class is not using the student notebooks, you will need to make copies of the record sheets and student instructions from the Teacher's Guide for your students, as directed in the individual lessons.

If you have these notebooks, you may want to collect them periodically to assess students' progress. At the conclusion of the unit, students may keep their notebooks as a reminder of all they have learned.

Teaching Strategies

Classroom Discussion: Class discussions, effectively led by the teacher, are important vehicles for science learning. Research shows that the way questions are asked, as well as the time allowed for responses, can contribute to the quality of the discussion.

When you ask questions, think about what you want to achieve in the ensuing discussion. For example, open-ended questions, for which there is no one right answer, will encourage students to give creative and thoughtful answers. You can use other types of questions to encourage students to see specific relationships and contrasts or to help them summarize and draw conclusions. It is good practice to mix these questions. It also is good practice always to give students "wait time" before expecting them to answer; this will encourage broader participation and more thoughtful answers. You will want to monitor responses, looking for additional situations that invite students to formulate hypotheses, make generalizations, and explain how they arrived at a conclusion.

Brainstorming: Brainstorming is a whole-class exercise in which students contribute their thoughts about a particular idea or problem. When used to introduce a new science topic, it can be a stimulating and productive exercise. It also is a useful and efficient way for the teacher to find out what students know and think about a topic. As students learn the rules for brainstorming, they will become increasingly adept in their participation.

To begin a brainstorming session, define for students the topics about which they will share ideas. Explain the following rules to students:

- Accept all ideas without judgment.
- Do not criticize or make unnecessary comments about the contributions of others.
- Try to connect your ideas to the ideas of others.

Cooperative Learning Groups: One of the best ways to teach hands-on science is to arrange students in small groups. In *Comparing and Measuring*, some materials and procedures are based on groups of four. There are several advantages to this organization. It provides a small forum in which students can express their ideas and get feedback. It also offers students a chance to learn from each other by sharing ideas, discoveries, and skills. With coaching, students can develop important interpersonal skills that will serve them well in all aspects of life. As students work, they will often find it productive to talk about what they are doing, resulting in a steady hum of conversation. If you or others in the school are accustomed to a quiet room, this new, busy atmosphere may require some adjustment.

Learning Centers: You can give supplemental science materials a permanent home in the classroom in a spot designated as the learning center. Students can use the center in a number of ways: as an "on your own" project center, as an observation post, as a trade-book reading nook, or simply as a place to spend unscheduled time when assignments are done. To keep interest in the center high, change the learning center or add to it often. Here are a few suggestions of items to include:

- Trade books on comparing and measuring and relative size (see the Bibliography for suggested titles).
- Measuring tools such as rulers, yardsticks, and tape measures so that students can practice measuring skills.
- Various objects of all sizes for students to measure.

Materials

Safety: This unit does not contain anything of a highly toxic nature, but common sense dictates that nothing be put in the mouth. In fact, it is good practice to tell your students that, in science class, materials are never tasted.

Organization of Materials: To help ensure an orderly progression through the unit, you will need to establish a system for storing and distributing materials. Being prepared is the key to success. Here are a few suggestions:

- Read through the Materials List on pg. 5. Begin to collect the items you will need that are not provided in the kit.
- Organize your students so that they are involved in distributing and returning materials. If you have an existing network of cooperative groups, delegate the responsibility to one member of each group.
- Organize a distribution center and instruct your students to pick up and return supplies to that area. A cafeteria-style approach works especially well when there are large numbers of items to distribute.
- Look at each lesson ahead of time. Some have specific suggestions for handling materials needed that day.
- Minimize cleanup by providing each working group with a cleanup box. Students can put disposable materials into this box and clean off their tables at the end of each lesson.
- Management tips are provided throughout the unit. Look for this icon:

Assessment

Philosophy: In the Science and Technology for Children program, assessment is an ongoing, integral part of instruction. Because assessment emerges naturally from the activities in the lessons, students are assessed in the same manner in which they are taught. They may, for example, perform experiments, record their observations, or make oral presentations. Such assessments permit the examination of processes as well as of products, emphasizing what students know and can do.

The learning goals in STC units include a number of different science concepts, skills, and attitudes. Therefore, a number of different strategies are provided to help you assess and document your students' progress toward the goals (see Figure T-1 on pgs. 10–12). These strategies also will help you report to parents and appraise your own teaching. In addition, the assessments will enable your students to view their own progress, reflect on their learning, and formulate further questions for investigation and research.

Figure T-1 summarizes the goals and assessment strategies for this unit. The left-hand column lists the individual goals for the *Comparing and Measuring* unit and the lessons in which they are addressed. The right-hand column identifies lessons containing assessment sections to which you can turn for specific assessment strategies. These strategies are summarized as bulleted items.

Assessment Strategies: The assessment strategies in STC units fall into three categories: matched pre- and post-unit assessments, embedded assessments, and additional assessments.

The first lesson of each STC unit is a *pre-unit assessment* designed to give you information about what the whole class and individual students already know about the unit's topic and what they want to find out. It often includes a brainstorming session during which students share their thoughts about the topic through exploring one or two basic questions. In the *post-unit assessment* following the final lesson, the class revisits the pre-unit assessment questions, giving you two sets of comparable data that indicate students' growth in knowledge and skills.

Throughout a unit, assessments are incorporated, or embedded, into lessons. These *embedded assessments* are activities that occur naturally within the context of both the individual lesson and the unit as a whole; they are often indistinguishable from instructional activities. By providing structured activities and guidelines for assessing students' progress and thinking, embedded assessments contribute to an ongoing, detailed profile of growth. In many STC units, the last lesson is an embedded assessment that challenges students to synthesize and apply concepts or skills from the unit.

Additional assessments can be used to determine students' understanding after the unit has been completed. In these assessments, students may work with materials to solve problems, conduct experiments, or interpret and organize data. In grades three through six, they may also complete self-assessments or paper-and-pencil tests. When you are selecting additional assessments, consider using more than one assessment to give students with different learning styles opportunities to express their knowledge and skills.

Documenting Student Performance: In STC units, assessment is based on your recorded observations, students' work products, and oral communication. All these documentation methods combine to give you a comprehensive picture of each student's growth.

Teachers' *observations and anecdotal notes* often provide the most useful information about students' understanding, especially in the early grades when some students are not yet writing their ideas fluently. Because it is important to document observations used for assessment, teachers frequently keep note cards, journals, or checklists. Many lessons include guidelines to help you focus your observations. The blackline master on pg. 13 provides a format you may want to use or adapt for recording observations. It includes this unit's goals for science concepts and skills.

Work products, which include both what students write and what they make, indicate students' progress toward the goals of the unit. Children produce a variety of written materials during a unit. Record sheets, which include written observations, drawings, graphs, tables, and charts, are an important part of all STC units. They provide evidence of each student's ability to collect, record, and process information. Students' science journals are another type of work product. In grades one and two, journal writings are primarily suggested as extension activities in many lessons. Often a rich source of information for assessment, these journal writings reveal students' thoughts, ideas, and questions over time.

Students' written work products should be kept together in folders to document learning over the course of the unit. When students refer back to their work from previous lessons, they can reflect on their learning. In some cases, students do not write or draw well enough for their products to be used for assessment purposes, but their experiences do contribute to their development of scientific literacy.

Oral communication—what students say formally and informally in class and in individual sessions with you—is a particularly useful way to learn what students know. This unit provides your students with many opportunities to share and discuss their own ideas, observations, and opinions. Some young children may be experiencing such activities for the first time. Encourage students to participate in discussions, and stress that there are no right or wrong responses. Creating an environment in which students feel secure expressing their own ideas can stimulate rich and diverse discussions.

Individual and group presentations can give you insights about the meanings your students have assigned to procedures and concepts and about their confidence in their learning. In fact, a student's verbal description of a chart, experiment, or graph is frequently more useful for assessment than the product or results. Questions posed by other students following presentations provide yet another opportunity for you to gather information. Ongoing records of discussions and presentations should be a part of your documentation of students' learning.

Figure T-1

Comparing and Measuring: Goals and Assessment Strategies

Concepts	
Goals	**Assessment Strategies**
Comparing involves observing similarities and differences. Lessons 1–16	Lessons 1–7, 11–12, 14–16, and Additional Assessments 2–3 ▪ Record sheets ▪ Body cutouts ▪ Representational graphs ▪ Teacher observations ▪ Class lists and charts ▪ Pre- and post-unit assessments
One way to make comparisons is by matching. Lessons 2, 4–6	Lessons 2, 4–6 ▪ Class discussions ▪ Representational graphs ▪ Teacher observations ▪ Record sheets ▪ Class charts
Using beginning and ending points and placing units end to end are important factors when measuring. Lessons 3–16	Lessons 1, 3–7, 11–12, 14–16, and Additional Assessment 3 ▪ Class discussions ▪ Representational graphs ▪ Teacher observations ▪ Class charts ▪ Record sheets ▪ Pre- and post-unit assessments
Nonstandard units of measure produce varying results. Lesson 7	Lesson 7 ▪ Class discussions ▪ Teacher observations ▪ Student products ▪ Class charts
Standard units of measure produce more consistent results than nonstandard units and make it possible to share information. Lessons 1, 8–16	Lesson 11–12, 14–16, and Additional Assessments 1–3 ▪ Class discussions ▪ Teacher observations ▪ Record sheets ▪ Class charts ▪ Student products ▪ Pre- and post-unit assessments
Different units and tools can be used to measure objects. Lessons 8–16	Lessons 11–12, 14, 16, and Additional Assessment 2 ▪ Record sheets ▪ Teacher observations ▪ Class discussions ▪ Class charts ▪ Student products
Long tools make it easier to measure long objects. Lessons 11–12, 15–16	Lessons 11–12, 16, and Additional Assessment 1 ▪ Class charts ▪ Record sheets ▪ Teacher observations ▪ Class discussions ▪ Student products
A common starting line is required to make fair comparisons. Lessons 4–16	Lessons 1, 3–7, 11–12, 14, 16, and Additional Assessments 1–3 ▪ Class discussions ▪ Student products ▪ Record sheets ▪ Teacher observations ▪ Pre- and post-unit assessments

Skills	
Goals	**Assessment Strategies**
Observing similarities and differences among objects. Lessons 1–16	Lessons 1, 3–7, 11–12, 14, 16, and Additional Assessments 1-3 ▪ Class discussions ▪ Student products ▪ Record sheets ▪ Teacher observations ▪ Pre- and post-unit assessments
Describing similarities and differences among objects. Lessons 1–16	Lessons 1, 3–7, 11–12, 14, 16, and Additional Assessments 1–3 ▪ Class discussions ▪ Record sheets ▪ Teacher observations ▪ Class charts ▪ Pre- and post-unit assessments
Placing objects in serial order on the basis of height or length. Lessons 2–4	Lessons 3–4 ▪ Representational graphs ▪ Teacher observations ▪ Class charts
Communicating observations, ideas, and questions through discussion, drawing, and writing. Lessons 1–16	Lessons 1, 3–7, 11–12, 14, 16, and Additional Assessments 1–3 ▪ Student products ▪ Class discussions ▪ Teacher observations ▪ Record sheets ▪ Pre- and post-unit assessments
Organizing information on representational graphs and charts. Lessons 3–4, 6–7, 9	Lessons 3–4, 6 ▪ Representational graphs ▪ Record sheets ▪ Class charts
Making predictions about the relative lengths and sizes of objects. Lessons 4–6, 8, 11–16	Lessons 4–5, 11, 14 ▪ Class discussions ▪ Teacher observations ▪ Record sheets ▪ Class charts
Using standard and nonstandard units of measure. Lessons 7–16	Lessons 7, 11–12, 14 ▪ Representational graphs ▪ Class discussions ▪ Teacher observations
Using groups of tens to quantify large numbers of units. Lessons 10–16	Lessons 11–12, 14, 16, and Additional Assessment 3 ▪ Teacher observations ▪ Class discussion ▪ Class charts ▪ Record sheets
Measuring using beginning and ending points. Lessons 8–16	Lessons 11–12, 14, and Additional Assessments 2–3 ▪ Class discussions ▪ Teacher observations

Figure T-1: *Comparing and Measuring:* Goals and Assessment Strategies *(continued)*

Attitudes	
Goals	**Assessment Strategies**
Interpreting results of measurements. Lessons 8–13, 15–16	Lessons 11–12, 16 ▪ Class discussions ▪ Class charts
Developing an awareness of self and others by comparing height, length of arms and legs, and body cutouts. Lessons 1–4	Lessons 1–4 ▪ Student products ▪ Class discussions ▪ Teacher observations ▪ Record sheets ▪ Class charts and lists ▪ Pre- and post-unit assessments
Developing an appreciation of the usefulness of measuring in our daily lives. Lessons 1–16	Lessons 1, 3–7, 11–12, 14, 16 ▪ Student products ▪ Class discussions ▪ Teacher observations ▪ Record sheets ▪ Class charts ▪ Post-unit assessment
Becoming comfortable using a variety of measuring tools and units of measure. Lessons 1–16	Lessons 1, 3–7, 11–12, 14, 16 ▪ Student products ▪ Class discussions ▪ Teacher observations ▪ Record sheets ▪ Class charts ▪ Pre- and post-unit assessments
Recognizing the importance of developing strategies for counting large numbers. Lessons 9–16	Lessons 11–12, 14, 16 ▪ Class discussions ▪ Teacher observations ▪ Record sheets ▪ Class charts
Appreciating the importance of organizing information on graphs and charts. Lessons 3–4, 6–7, 9	Lessons 3–4, 6–7 ▪ Representational graphs ▪ Class discussions

Blackline Master

Comparing and Measuring: Observations of Student Performance

STUDENT'S NAME:

Concepts	**Observations**
• Comparing involves observing similarities and differences.	
• One way to make comparisons is by matching.	
• Using beginning and ending points and placing units end to end are important factors when measuring.	
• Nonstandard units of measure produce varying results.	
• Standard units of measure produce more consistent results than standard units and make it possible to share information.	
• Different units and tools can be used to measure objects.	
• Long tools make it easier to measure long objects.	
• A common starting line is required to make fair comparisons.	
Skills	
• Observing similarities and differences among objects.	
• Describing similarities and differences among objects.	
• Placing objects in serial order on the basis of height or length.	
• Communicating observations, ideas, and questions through discussion, drawing, and writing.	
• Organizing information on representational graphs and charts.	
• Making predictions about the relative lengths and sizes of objects.	
• Using standard and nonstandard units of measure.	
• Using groups of tens to quantify large numbers of units.	
• Measuring using beginning and ending points.	
• Interpreting results of measurements.	

LESSON 1 Comparing How We Are Alike and Different

Overview and Objectives

In this first lesson, students are introduced to comparing as a way to observe similarities and differences. Students describe ways they are alike and different and have the opportunity to use simple tools such as coffee stirrers, pencils, and spoons to aid in measuring and making comparisons. Both activities will provide assessment information on students' ability to observe, compare, and communicate ideas. Students' written observations and the ideas they suggest when compiling a class list give you further information about the specific comparisons they make. By contrasting this information with parallel information at the end of the unit, you will be able to assess the growth in each student's ability to compare and measure and the methods they use for each. The activities in this lesson set the stage for Lesson 2, where students will use life-size cutouts of their bodies to continue exploring ways of comparing.

- Students discuss what it means to compare and measure.
- Students observe each other and identify similarities and differences.
- Students discuss their comparisons.
- Students record their observations.
- Students have the opportunity to make comparisons using various everyday objects such as coffee stirrers, pencils, and spoons.

Background

Why have a science unit on comparing and measuring? When scientists do experiments, they often need to measure; that is, using numbers and standard units of measure, they describe such properties as length, volume, weight, and temperature. Similarly, when students do experiments, they will need skills in measuring these properties.

Even outside of the classroom, students frequently measure and compare; for example, when they make kites, wrap presents, or chart their heights. In fact, measuring and comparing are key means through which students make sense of their lives. In this unit, students will gain experience in using measuring tools to obtain information and to solve problems. They will follow the developmental sequence of comparing, matching, and measuring as they observe similarities and differences and match and measure lengths, heights, and distances.

Comparing is the first step in the developmental sequence of attaching meaning to a numerical measurement. When a child yells, "I jumped farther than Joseph," he or she is making a visual estimate of the two distances jumped and comparing them. To confirm if one jump was truly farther than the other jump, the child could **match** each distance to a piece of rope and compare the lengths of the two pieces of rope. However, if the child wanted to numerically quantify how much farther his or her jump was, the student would need to measure the distance of both jumps and compare the measurements to get the difference.

In this lesson, students begin by brainstorming what they already know about comparing and measuring. They then observe the differences and similarities among themselves. Some students may want to use materials in the classroom or distribution center to help them make comparisons. For example, a student may use toothpicks and say, "My arm is eight toothpicks long." Another student may not use numbers and standard units but simply match his arm against the length of his classmate's and say, "my arm is longer than yours". Or those students who have worked with Unifix Cubes™ before may know to hook the cubes together and use them to assign a specific value to the length of an object.

This is an unstructured activity; you will leave it up to students as to what—if any—materials from the distribution center they will use in making their comparisons. By not directing students toward any specific use of materials, you achieve baseline assessment information on their comparing and measuring abilities.

Many students' initial observations may focus on similarities and differences in physical appearance, such as hair or skin color, types of clothes or shoes, and gender. Depending partly on their developmental level, some students may make linear comparisons, such as indicating who is taller or shorter. At the end of the unit, you will compare students' observations in this lesson with those in the post-unit assessment. At that time, you can expect students' comparisons to be more diverse and to demonstrate what students have experienced in the course of the unit.

Materials

For each student

- 1 copy of **Record Sheet 1-A: Looking at My Partner and Me**
- 1 package of crayons, including one red crayon and one blue crayon
- 1 resealable plastic bag with which to collect materials, 23 × 30 cm (9 × 12 in)
- 1 scissors

For the class

- 2 sheets of newsprint
- 1 marker
- 1,500 Unifix Cubes™, separated by color
- 1 container of each of the following:
  - 100 wood coffee stirrers
  - 100 unsharpened pencils
  - 100 plastic spoons
  - 100 toothpicks
  - 100 small wood spools, 4 cm (1½ in)
- 15 rolls of adding machine tape

Notes: *My Comparing and Measuring Book,* the student notebook, contains all the record sheets and reading selections used in this unit. If your class is not using the student notebook, you will need to photocopy these pages for your students. All items that you must photocopy appear in the teacher's guide at the end of the lesson in which they will be used.

Materials from the distribution center can be collected and returned in the resealable bags. This procedure can be followed throughout the unit. Students can also use shoe boxes, foam meat trays, or plastic jugs in which to collect materials.

Preparation

1. Label a sheet of newsprint "What We Know about Comparing and Measuring." Write today's date and lesson number on it.
2. Label the second sheet of newsprint "Ways We Are Alike and Different" and write the date on it.

Management Tip: Throughout this unit, students generate many charts and lists that provide a record of the discoveries they are making. To make it easier to refer back to a specific chart for assessment purposes, teachers have found it helpful to record the date and lesson number on each chart. If possible, display the charts in the classroom throughout the unit so that students may refer to them. If space is a problem, you may consider rotating the charts, hanging them on a wire "clothesline" in the classroom, or binding them together into a "Big Book."

3. Set up a distribution center in your classroom (see Figure 1-1). You will need to find containers such as shoe boxes, file folder boxes, or plastic tubs to hold the materials.

   Following are the steps for setting up a distribution center:

   - Select one large area of the room.
   - Place all of the materials on a large table, windowsill, or several desks that have been pushed together.
   - Arrange the materials in separate containers. Before each lesson, place a label on each container indicating what it holds. (In this lesson, students will take as many of each material as they want. In subsequent lessons, the label will also need to indicate how many of each item the students should take.) Make sure students have enough space to walk by when gathering materials.

Figure 1-1

Distribution center

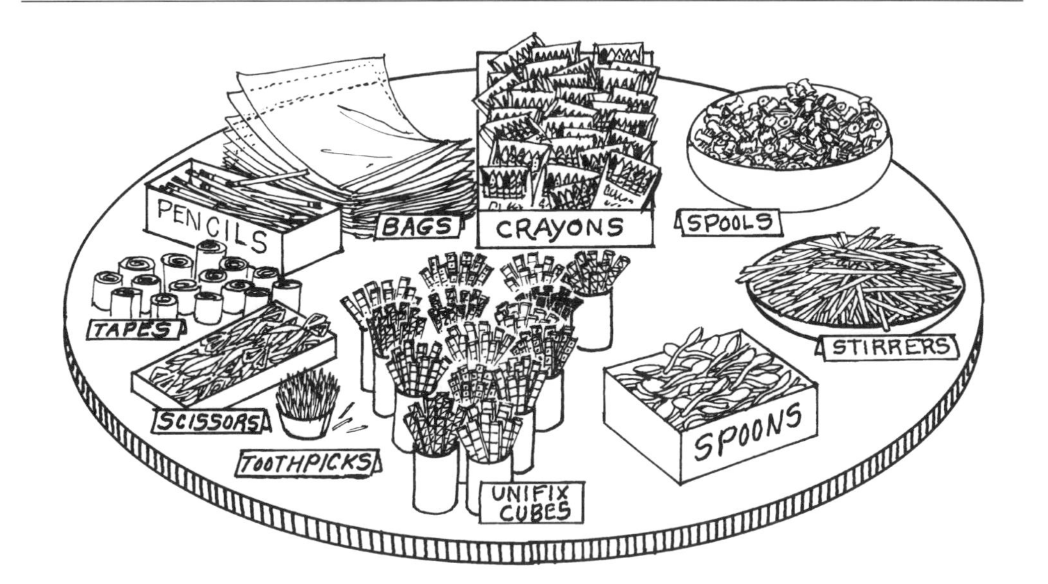

4. Arrange students in pairs.
5. Copy **Record Sheet 1-A: Looking at My Partner and Me** for each student.

Procedure

1. Introduce students to this unit by letting them know that for the next eight weeks, they will be exploring comparing and measuring.
2. Ask students to think about what they know about comparing and measuring. After a few minutes, have them share their thoughts with the class. Record these thoughts on the "What We Know about Comparing and Measuring" chart. To help stimulate student discussion, you may want to ask questions such as the following:
   - When have you compared before? When have you measured before?
   - How did you compare? How did you measure?
   - Why were you comparing? Why were you measuring?

   Save this chart for use in the post-unit assessment.
3. Display the "Ways We Are Alike and Different" chart. Let students know you would like partners to decide on one way they are like each other and one way they are different from each other. Let students know that when they look at how things are alike and different, they are comparing. Invite students to use any materials in the classroom or distribution center to help them find out about their partners.
4. After a few moments, ask students to share their thoughts. To encourage discussion, ask the class questions such as the following:
   - In what way are you and your partner alike?
   - How are you and your partner different?
   - Did you use any materials from the distribution center to help make your comparisons?
   - How did these materials help you make comparisons?
5. Record students' thoughts on the chart (see Figure 1-2).
6. Leave the chart on display for use in the next lesson.

Final Activities

1. Pass out and review **Record Sheet 1-A: Looking at My Partner and Me.** Ask students to do the following:
   - Write your name and today's date on the record sheet.
   - Draw a picture of yourself and your partner. Write your partner's name in the box with his or her picture.
   - Draw a red circle around the part of the picture that shows one way you and your partner are alike.
   - Draw a blue circle around the part of the picture that shows one way you and your partner are different.
   - Write one or two sentences describing each likeness and difference.
2. On the chalkboard, you may want to write sentence starters such as the following:
   - I am like my partner because ________________________.

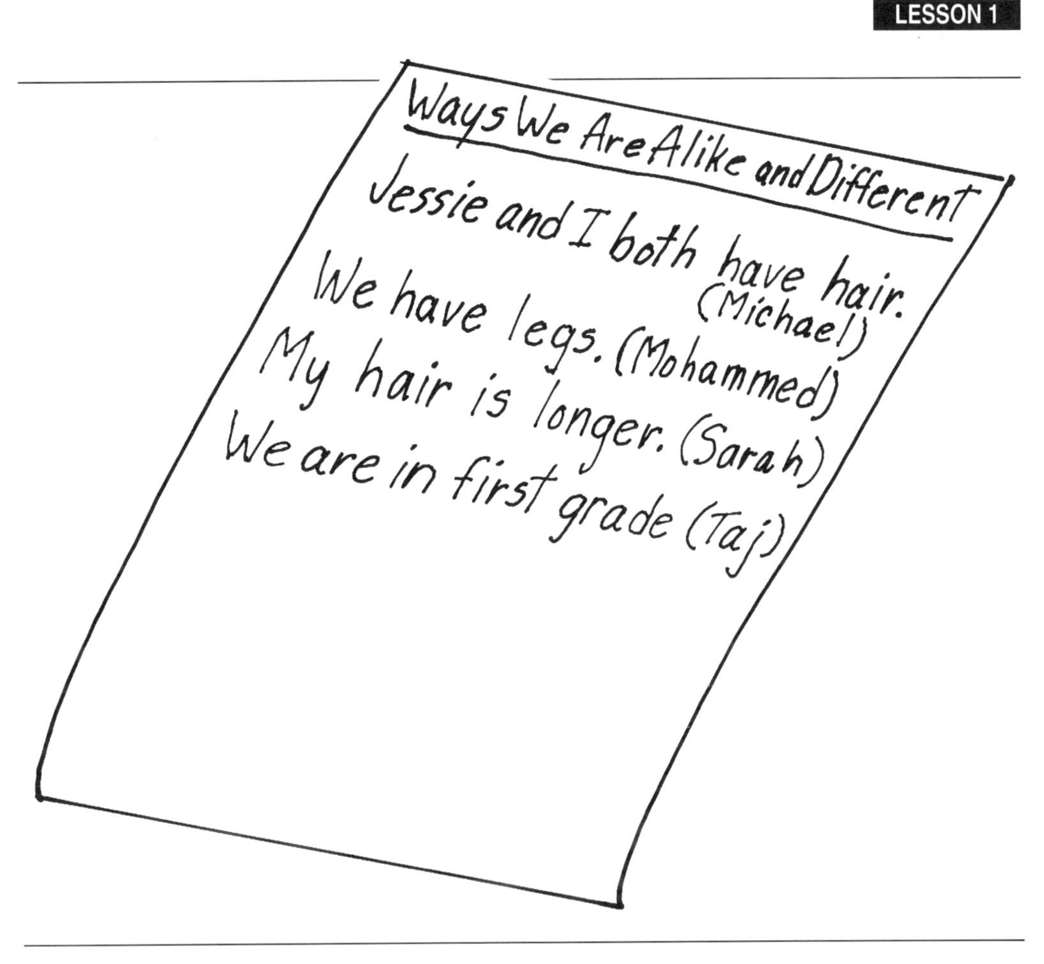

Figure 1-2

Sample chart of students' ideas

- One way I am different from my partner is ______________________.
- My partner and I ______________________.

3. Invite students to share and discuss their drawings with the class.
4. Collect the record sheets and save them for use in the post-unit assessment.

Extensions

LANGUAGE ARTS | MATHEMATICS

1. Ask students to look through old magazines and catalogs for pictures of objects such as cars, animals, food, and families. They can cut the pictures out and group like objects together. Students can glue the pictures to newsprint and label and discuss their ideas.

LANGUAGE ARTS | SOCIAL STUDIES

2. Have children bring in pictures of the people in their families and share ways the families are alike and different.

LANGUAGE ARTS

3. Read *People*, by Peter Spier (see Bibliography).

MATHEMATICS

4. Begin a height chart for each child and record his or her height throughout the year.

Assessment

In the section entitled Teaching *Comparing and Measuring* on pgs. 7–13, you will find a detailed discussion about the assessment of students' learning. Specific goals and related assessments for this unit are summarized in Figure T-1, *Comparing and Measuring*: Goals and Assessment Strategies, on pgs. 10–12.

The brainstorming chart "Ways We Are Alike and Different" is the first half of a matched pre- and post-unit assessment. The second half appears in the post-unit assessment on pgs. 129–31. Both are important components of your assessment of students' growth and learning.

Students' writings and drawings about what they observe and discover can provide evidence of the progression of their ideas. These materials will also alert you to students' questions and give you insights into topics in which they are especially interested. You may want to review your students' notebooks periodically or conduct individual science meetings that give each student an opportunity to share some of the ideas he or she has written about. You may also want to keep an observation log to aid in the assessment process throughout the unit.

To assess students in this lesson, look at their notebooks and observe students during the activities. As you do so, ask yourself the following questions:

- How do students define comparing and measuring?
- What similarities do students observe and describe?
- What differences do students observe and describe?
- Do students describe any differences that are linear, such as differences in height?
- Do record sheet entries demonstrate students' ability to distinguish between similarities and differences?
- If students used materials from the classroom or distribution center, how did they use them to make comparisons?

Record Sheet 1–A

Name:

Date:

Looking at My Partner and Me

My partner	Me

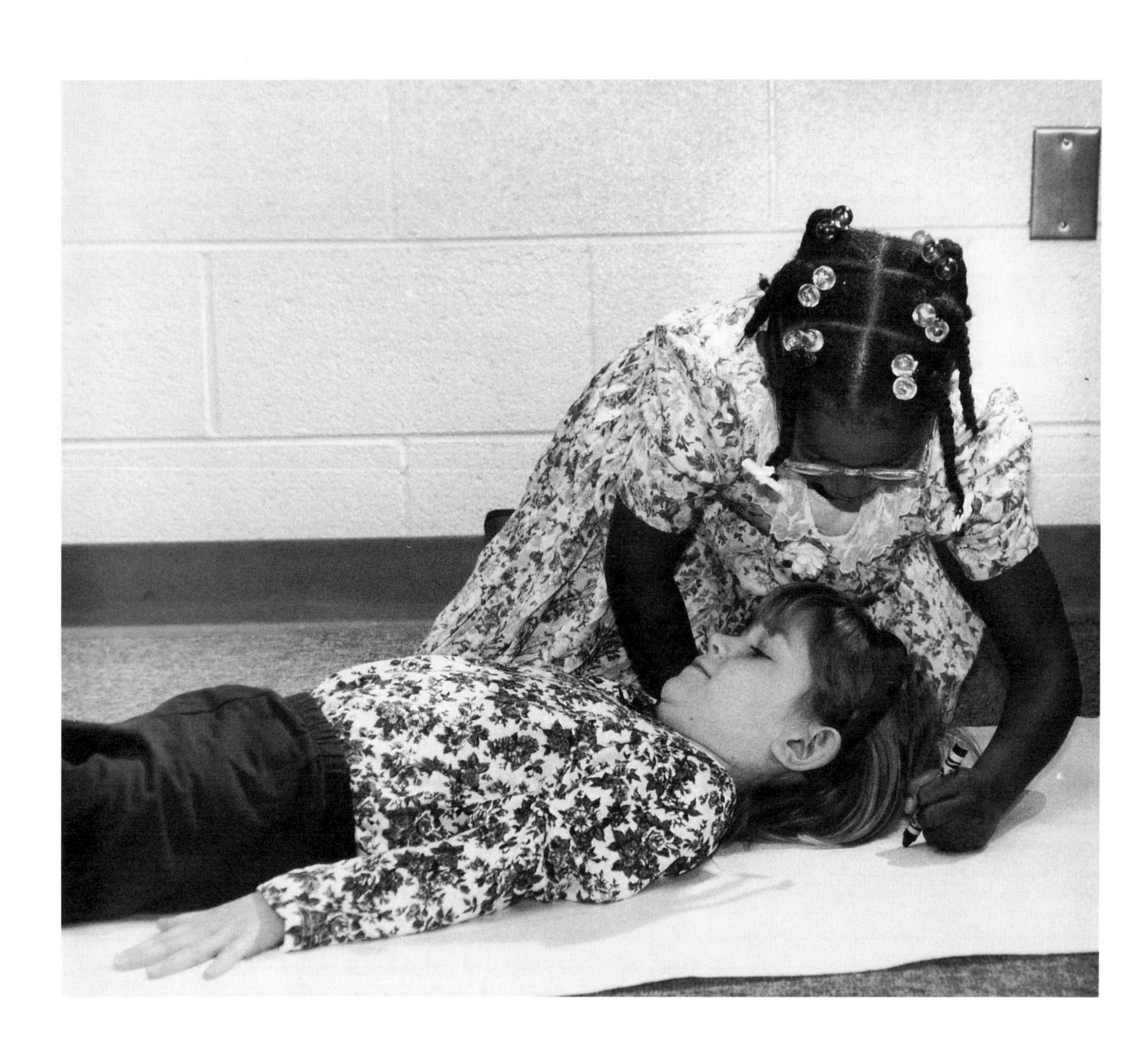

LESSON 2

Let's Make Body Cutouts

Overview and Objectives

In this lesson, students build on their experiences in describing ways they are alike and different. Students continue to recognize that they can use 'tools' other than themselves to match and make comparisons. They make life-size cutouts of their bodies and use these cutouts to continue observing and describing similarities and differences among themselves. Students are also introduced to the idea that the cutouts can be used to compare height. After placing the cutouts in serial order by height, students begin using a form of representational graphing that they will continue to use throughout the unit. This lesson prepares students for matching their heights with adding machine tape in Lesson 3.

- Students trace each other's body outlines.
- Students cut out the body outlines.
- Students compare similarities and differences among the cutouts.
- Students arrange the cutouts in serial order from shortest to tallest.

Background

By creating the body cutouts in this lesson, students are introduced to the concept of **matching;** that is, using one object to represent the length of another object. The outlines that students trace of their bodies represent the size of their physical bodies. The cutouts help students begin to recognize that they can use a representation of an object, rather than the object itself, to make comparisons.

As children use the body cutouts to make comparisons, you can expect them to continue observing similarities and differences in their physical appearances. You may notice some students beginning to make linear comparisons. For example, two students who hold their cutouts side by side may observe that one is taller, shorter, or the same height as the other. To encourage students to make this type of observation, you will ask them to arrange the cutouts in order from shortest to tallest and to explain how they arrived at their orderings. To do so, they will actually be analyzing their own data—an important science skill. Students will begin to recognize that they must organize, interpret, and apply the data they collect through scientific investigations to make that data meaningful.

Materials

For each student

- 1 sheet of large paper for tracing the body outline
- 1 marker
- 1 pair of scissors
- 1 pencil
- Crayons

For the class

- 1 sheet of newsprint
- 1 marker
- 1 "Ways We Are Alike and Different" chart (from Lesson 1)

Preparation

1. Cut out one large sheet of paper (bulletin board or butcher paper works well) for each student. The sheets should be long enough so that a student can lie down on it with no body parts extending off the paper (see Figure 2-1). You may want to find your tallest student and cut the paper to fit his or her height. The sheets will then be long enough to accommodate the other students.

Figure 2-1

Tracing body outlines

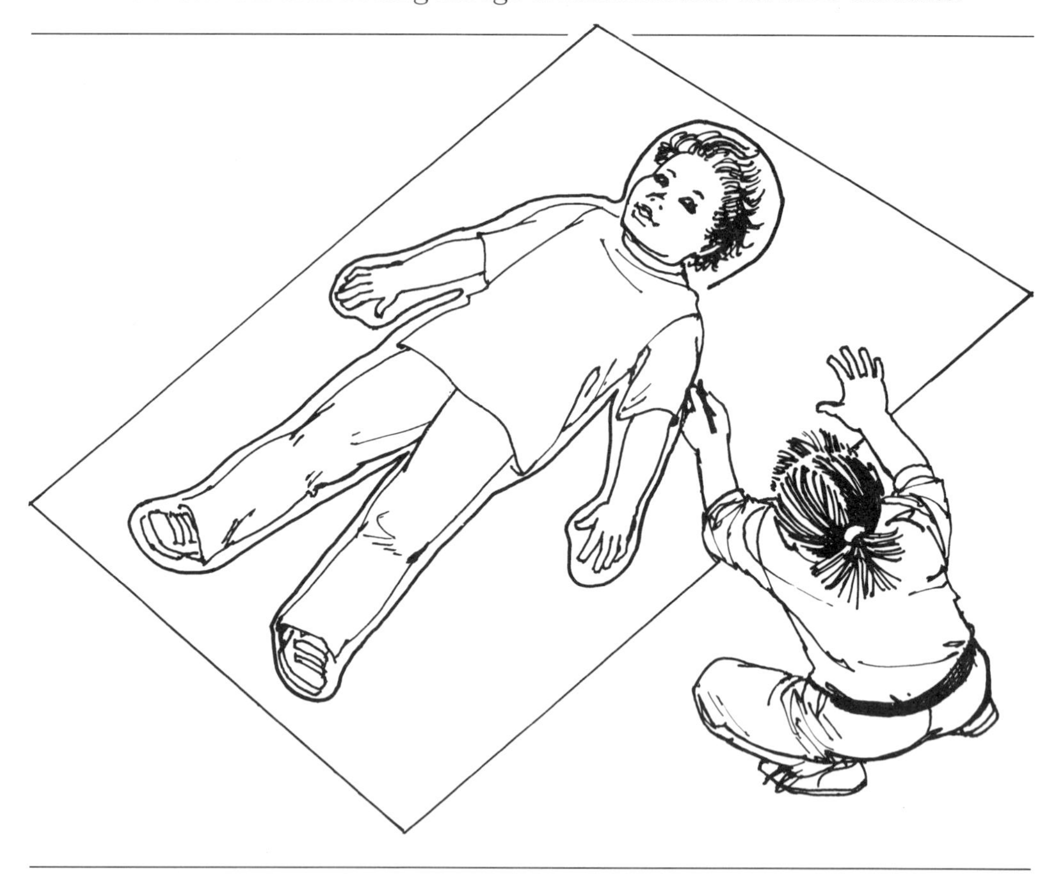

2. On the sheet of newsprint, write the title "Ways to Compare Our Bodies" and the date.
3. Arrange students in pairs.

   Note: When arranging students in pairs or groups, consider mixing students of different ability levels. That way, each pair or group has strengths in different areas, and students are likely to complement each other and work well together.

4. Make sure the "Ways We Are Alike and Different" chart from Lesson 1 is on display.

Procedure

1. Have students focus on the "Ways We Are Alike and Different" chart from Lesson 1 and review the ideas they brainstormed. Encourage students to identify any words on the chart that show they made comparisons about size (for example, shorter, taller, longer, or smaller). Have volunteers use a marker or crayon to circle these words on the chart.

2. Now ask students to think about how they could compare their bodies with those of their partners. Let students take time to explore their ideas with their partners and then have them share their ideas with the class. Record these ideas on the "Ways to Compare Our Bodies" chart. In the past, students have said things such as, "We compared our hands and feet by holding them against each other" and "Sara and I stood back to back."

3. Let students know that another way to compare their bodies is to make body cutouts. Describe the process students will follow to make body cutouts of their partners:

   - With your partner, collect the materials from the distribution center.
   - Have your partner lie on his or her back on the piece of paper. Make certain that no part of your partner's body is off the paper.
   - With a dark crayon, closely trace around your partner's body, including hair, fingers, and clothing. Your partner should lie on the paper without moving.
   - Cut out the body outline. Color it if you have time.

Management Tip: You may want to arrange for a "buddy" class of fourth- or fifth-graders to assist your students should they need help tracing, cutting, or coloring the body cutouts.

4. As the students are tracing, circulate to be sure that their entire bodies are on the sheets of paper. Students can include clothing and facial features on the cutouts if there is enough time. If not, extend the lesson to the next day.

Final Activities

1. Have students clean up their materials. Bring the class together. Be sure students have their body cutouts.

   Note: If your classroom is not large enough, find a space such as the hallway or lunchroom where all the cutouts can be displayed and observed at the same time.

2. To help your students verbalize their observations, you may want to ask some of the following questions:

   - How would you describe your body cutout and compare it with your partner's?
   - Using your body cutouts, how could you and your partner find out who is taller and who is shorter?

3. Ask students to look at the body cutouts of others in the class. Invite them to find one cutout that is in some way like their own and one cutout that is in some way different from their own.

4. Have students share their observations with the class. Record them on the "Ways We Are Alike and Different" chart. Date today's observations and use a different color to record them. Display the chart in the classroom and save it for the post-unit assessment.

5. Now arrange students in groups of four. Ask each group to put the body cutouts in order from shortest to tallest. Ask students how they decided upon the order of the cutouts.

6. Save the cutouts and display them if possible. They may be used for the additional assessment on pg. 135. As students explore other strategies for matching and measuring in this unit, they may want to use machine tape, coffee stirrers, or Unifix Cubes™ to compare the sizes.

Extensions

LANGUAGE ARTS

1. Read *A Big Fish*, by Joanne and David Wylie (see Bibliography).

LANGUAGE ARTS

2. Have students think and write about what would happen if their body cutouts came to life and ran away. To encourage ideas, ask the following questions:

   - What do you think your body cutout would want to do?
   - Where do you think your body cutout would want to go?
   - Whom would your body cutout talk to and what would it say?

3. Use a cookie cutter shaped like a body to make cookies. Have students decorate the cookies in different ways and talk about how they are alike and different.

ART

4. Have children make body cutouts of the principal, secretary, or janitor and add hair and clothing. Students can compare their own body cutouts with those of the new people.

Comparing Heights

LESSON 3

Matching Our Heights

Overview and Objectives

In the last lesson, students used their representational body cutouts to compare likenesses and differences. Now they represent their heights with adding machine tape and use the tape to compare and seriate their heights. By matching the tape to their heights, students begin to see that when matching the tape to a person or object, they have to determine both a beginning point and an ending point.

- Students match their heights with adding machine tape.
- Students arrange the tapes in serial order from shortest to longest on a representational graph.
- Students discuss how they determined beginning and ending points.

Background

An important, yet difficult, skill underlying matching and measuring is determining beginning and ending points. When students created the body cutouts in Lesson 2, they automatically determined beginning and ending points. Simply by tracing around their bodies, they determined the limits of what was being matched. In this lesson, they must decide for themselves where, in relationship to their bodies, to place the adding machine tape and where to cut it in order to make strips that represent their heights.

Most students will have had the experience of having their height marked while standing next to a wall or of having their height measured at the doctor's office. Because of these experiences, many students will use the top of the head as the beginning point for the tape. However, they may not have a sense of where to end the tape. For example, students may stop at the ankles, roll the tape past the feet, or pay no attention whatsoever to the position of the feet. These variables will affect how long the students make the tapes. This, in turn, will affect how accurately the tapes represent height. Students discuss these variables when they examine the serial order created by arranging the tapes from shortest to longest on the graph.

When constructing the graph of their heights, students may not recognize the need for a common starting line. This line is the point where the tapes begin and enables students to make accurate comparisons among tapes of different lengths.

Materials

For each student

1 roll of adding machine tape
1 bottle of glue

For every two students

1 pair of scissors
1 marker

For every four students

2 sheets of newsprint

For the class

1 sheet of newsprint

Preparation

1. Arrange materials for this lesson in the distribution center.
2. On a sheet of newsprint, write the title "Being Measured" and the date. Display the chart.
3. For each group of four students, tape two sheets of newsprint end to end so that you have one long sheet. Label the top of the sheet "Our Heights," as shown in Figure 3-1.
4. Arrange your class into teams of four. Students will work in these same teams in Lesson 4. Within each team, make two pairs of partners.

Figure 3-1

Making the "Our Heights" graph

Procedure

1. To begin the lesson, ask students the following questions:
   - Can you think of a time that someone needed to find out how tall you were?
   - Where were you?
   - Why did the person need to find out your height?
   - What did the person use to measure you?
   - Record students' responses on the "Being Measured" chart.
2. Discuss with students that in today's lesson, they will collect information about their heights and record it on a graph.
3. Describe the process students will follow.
   - With your partner, collect the materials from the distribution center.
   - Have your partner lie on his or her back on the floor.
   - Use the adding machine tape to match your partner's height.
   - Cut the tape and write your partner's name (the name of the student being measured) on the tape.
   - Switch places and follow the same procedure.
4. After all members of the team have used the tape to match their partners' heights, have one student from each group get one of the charts labeled "Our Heights" from the distribution center. Then ask students to do the following:
   - Place the four tapes side by side on the floor and compare heights.
   - Arrange the tapes in order from shortest to tallest.
   - With your team members, check the tapes again to see if you might need to change the order.
   - When the team has agreed upon the order of the tapes, glue them to the graph in that order. Make sure your names on the tapes are easy to see and read.
   - Collect all the materials and return them to the distribution center.
5. Circulate as students are matching with the tape and constructing the graph.

Final Activities

1. Give the teams a few minutes to observe their graphs. Ask them to talk with each other about the information the chart provides about the team members' heights. Have students think about the following questions:
   - How did you know where to begin measuring your partner's body?
   - How did you know where to end measuring your partner's body?
   - Did everyone on your team use the same beginning and ending points on their partners?
   - Why is it important to use the same beginning and ending points?
2. Have each team display its graph and share information such as which tapes are the longest, the shortest, and about the same length. Ask students to think of and discuss comparisons among the height graphs. Previously, student responses have included "Some groups had more tall people than others," "The tapes are all different lengths," and "One group glued their tapes from the top down and other groups didn't." Students may not recognize the need for a common starting line. The importance of using a common starting line will be addressed more fully in Lesson 4.

3. Ask the students, "What do you now know about yourself after using the adding machine tape?"
4. Display the graphs in the classroom. Save them for use in Lesson 4.

Extensions

LANGUAGE ARTS

1. Write a "tall" book. Cut unlined paper into 15-by-28-cm (6-by-11-in) strips. Cut construction paper to the same dimensions as the unlined paper. To make a book, put six or seven paper strips between two pieces of construction paper and staple the paper together. Have students brainstorm some things that are tall. Record these items on a piece of chart paper. Have students write sentences about the tall things and illustrate their books.

MATHEMATICS

2. Have a "mystery tape box" in the class. Use the adding machine tape to match the students' heights. Number the tapes and record the students' names so you will know which tape matches each student's height. Each day, place a tape in the "mystery tape box" and have children match the tape to one another's heights and try to identify the student it matches.

LANGUAGE ARTS

3. Read *Once I Was Very Small*, by Elizabeth Ferber (see Bibliography).

Assessment

Lessons 3 through 5 provide an opportunity for you to assess students' progress in understanding the use of common starting lines and beginning and ending points. As you observe students matching their heights and discuss their ideas, consider some of the following questions:

- Do students recognize that they need to use a beginning point and an ending point when matching with the tape?
- Do students discuss the need to be consistent by using the same beginning and ending points each time they match height?
- Do students recognize whether the tape is clearly longer or shorter than the student?
- Do students suggest rematching to obtain a more accurate length of tape?
- Do students recognize the need for a common starting line as they compare results?

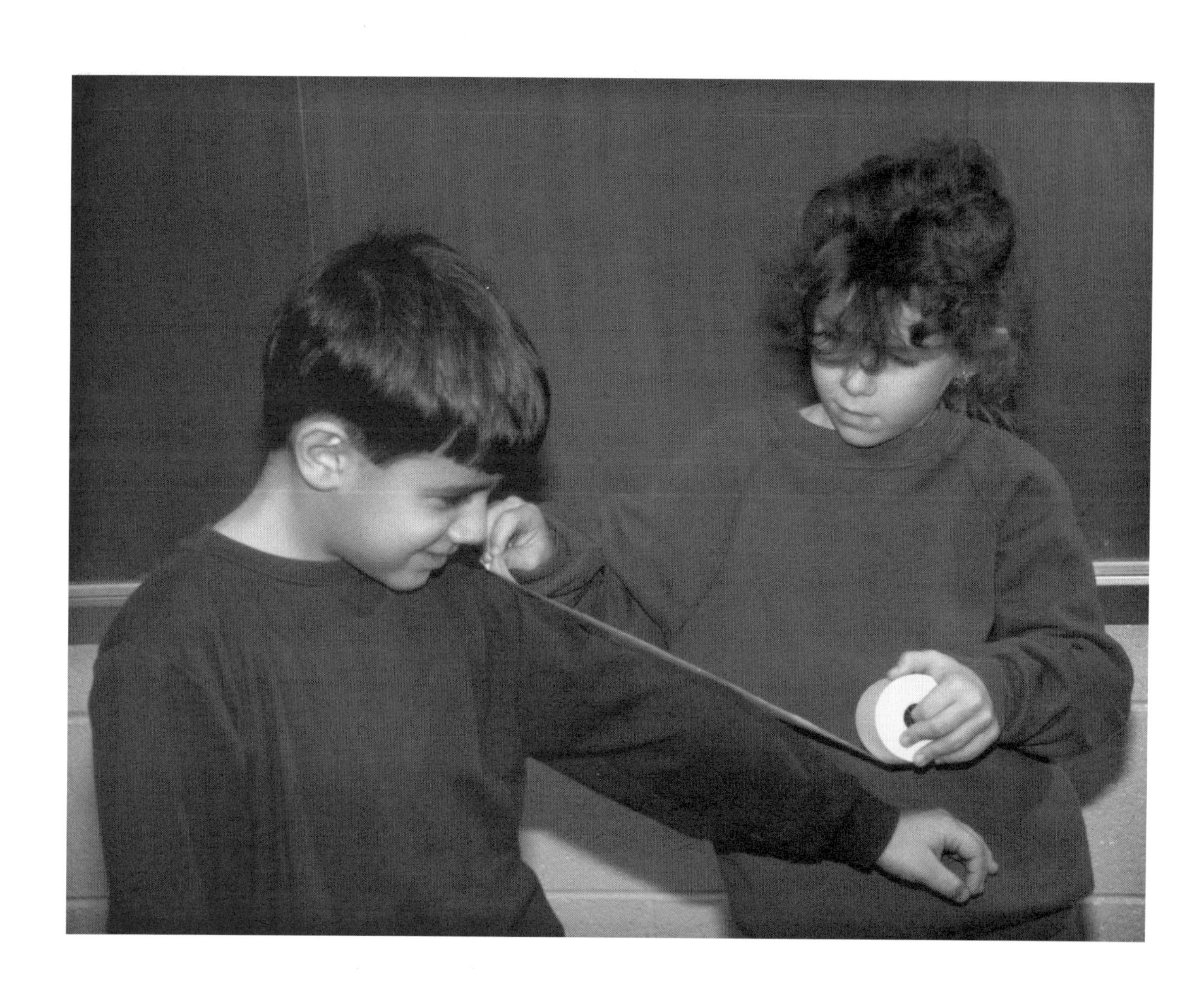

LESSON 4

Matching Lengths of Arms and Legs

Overview and Objectives

In this lesson, students have an opportunity to apply what they have learned in Lesson 3 about the importance of using beginning and ending points when matching. They build on this idea by exploring the use of a common starting line when comparing on a representational bar graph. Appreciating the need for a common starting line helps strengthen students' concept of making fair comparisons. In addition, students take the first step toward predicting by making a guess about the lengths of their arms and legs. Throughout the unit, students will gain more experience with using a common starting line and making predictions.

- Students guess which team member has the longest arm and longest leg.
- Students match the lengths of their arms and their legs and graph their results in serial order from shortest to longest.
- Students discuss why it is important to use a common starting line when making comparisons.
- Students compare the information on the arm and leg graphs with the information on the height graph from Lesson 3.

Background

When students match the lengths of their arms in this lesson, they will place the adding machine tape from the wrist to the end of the shoulder, as in Figure 4-1 When students match the lengths of their legs, they will place the tape from the outside of the ankle to the outside of the hip, as in Figure 4-2.

When students construct a graph in this lesson, they will begin to recognize the importance of a common starting line. You may want to explain that in a race, people need to begin, as well as end, at the same place. If everyone started from different places, they would be running different lengths, and the race would not be fair. Similarly, by placing the arm and leg tapes at a common starting line, the students will be able to make fair comparisons.

In this lesson, students also gain experience making guesses, the first step in learning to make predictions. You may want to mention to students that scientists often make predictions; for example, a meteorologist makes predictions about the coming weather.

Figure 4-1

Matching arms

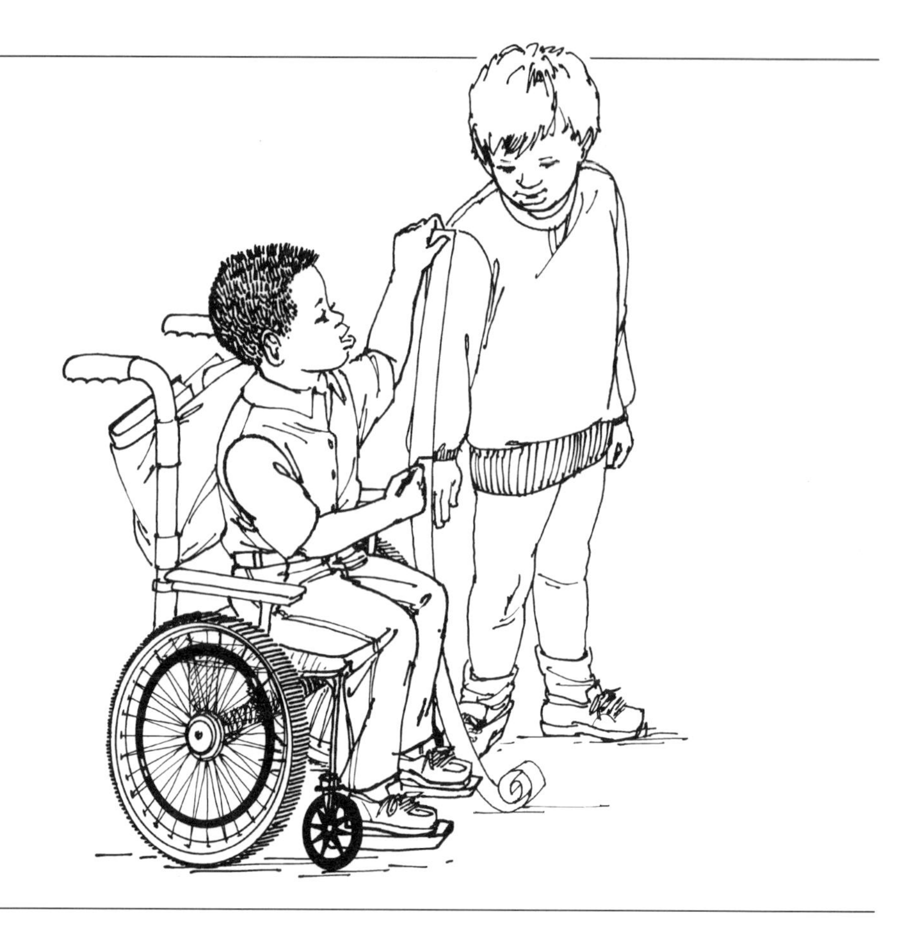

Figure 4-2

Matching legs

Materials

For each student

1 science journal
1 **Record Sheet 4-A: Matching Our Arms and Our Legs**

For every two students

1 roll of yellow adding machine tape
1 roll of pink adding machine tape
1 pair of scissors
1 marker
1 bottle of glue
1 resealable plastic bag, 23 × 30 cm (9 × 12 in)

For every four students

1 "Our Heights" graph (from Lesson 3)
1 sheet of newsprint

For the class

1 red marker

Preparation

1. Arrange the materials in the distribution center. Students will use the pink adding machine tape to match arms and the yellow to match legs. This will help them arrange and read the information on the graphs more easily.

   Note: This is the only lesson in which the color of adding machine tape is important. In subsequent lessons, students can choose to use either pink or yellow tape for matching.

2. Prepare a piece of newsprint for each team, as illustrated in Figure 4-3. Be sure to use a red marker to make the starting line. Label the graph "Arms/Legs."

Figure 4-3

Sample graph

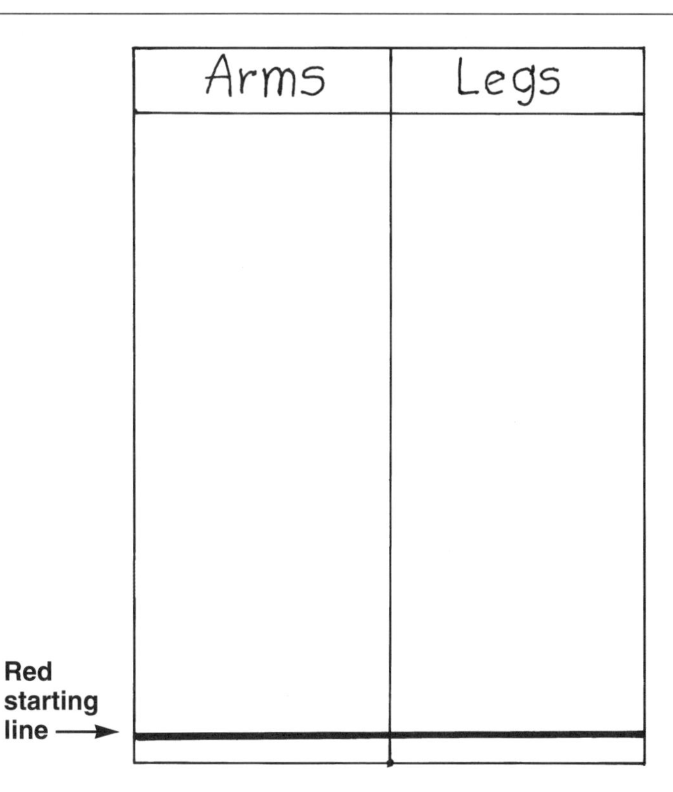

3. Group students in the teams from Lesson 3. Within each team, match partners.
4. Copy **Record Sheet 4-A: Matching Our Arms and Our Legs,** on pgs. 40–41, for each student.

Procedure

1. Have students prepare a science notebook (loose-leaf notebooks or folders with notebook paper inside work well) that they will use throughout the unit. Explain that the notebook will be a record of their thoughts, ideas, and drawings. Then ask students to think about some reasons for writing down the date above each day's entry.
2. Show the class the "Arms/Legs" graph. Ask students questions such as the following:
   - Why would someone need to know the length of your arms or legs?
   - If you were buying clothes, would you need to know the length of your arms and legs? Why?
3. Share with students that today they will gather information about the lengths of their arms and legs. They will record it on a graph, as they did in Lesson 3.

   Note: You can decide which arm and leg the teams will match (right or left), or you can let each team decide. It may help avoid confusion if you establish the beginning and ending points on the arm (wrist and shoulder) and leg (ankle and hip) before the students move to the next step.
4. Distribute **Record Sheet 4-A: Matching Our Arms and Our Legs** to each student.
5. Describe the process students will follow.
   - With your partner, collect the materials from the distribution center.
   - Everyone on the team should match the same arm (right or left).
   - Without actually placing your arms side by side, simply observe your arm and compare its length with the lengths of the arms of your team members.
   - Make an individual guess about which team member has the longest arm. Record your guess on Record Sheet 4-A.
   - Match the pink tape to the length of your partner's arm. Match the tape from your partner's wrist to his or her shoulder. Cut or tear the tape and write your partner's name on it.
6. Ask students to switch roles so that both partners have had an arm matched with the tape. Then invite students to do the following:
   - Everyone on the team should match the same leg (right or left).
   - Without actually placing your legs side by side, simply observe your leg and compare its length with the lengths of the legs of your team members.
   - Now make an individual guess about which team member has the longest leg. Record your guess on Record Sheet 4-A.
   - Using the yellow adding machine tape, match your partner's leg from ankle to hip. Cut or tear the tape and write your partner's name on it.
7. Pass out an "Arms/Legs" graph to each team. Explain the process for constructing the graph.
   - Using the red line as the common starting line, arrange your team's arm tapes from shortest to longest on the section of the graph labeled "Arms."

- Using the red line as the common starting line, arrange your team's leg tapes from shortest to longest on the section of the graph labeled "Legs."
- After your team has checked the tapes to see that they are in the right order, glue them to the graph.
- After comparing lengths of arms and legs, write the name of the team member with the longest arm on your record sheet. Do the same for the team member with the longest leg.

8. Circulate to assess how the students are comparing and arranging the tapes. You may want to record any observations in a log to aid in the assessment process throughout the unit.

Final Activities

1. Give the teams a few minutes to observe and discuss their graphs.
2. Have each team share its graph with the class.
3. Ask students to compare their "Arms/Legs" graphs with the "Our Heights" graphs from Lesson 3 and find ways that they are alike and different. Students have said, "We used adding machine tape in both graphs," "All of our tapes are sitting on the red line," and "People glued the height tapes all over."
4. Ask students to discuss the following questions:
   - How did you use the red starting line on the graph?
   - Is it important to have a common starting line? If yes, why?
   - How is this graph like your height graph? How is it different?
   - What information can you read from the graph?
5. Have students return their materials to the distribution center.

Extensions

1. Ask parents or an alterations shop to donate a few clothing patterns that are no longer used. Explain to students how the patterns were used to make clothes. Have students use newspaper, craft paper, or construction paper to make clothing for their body cutouts.

MATHEMATICS

2. Have your students match the arms and legs of other first-graders and older and younger students. Students can then compare these tapes with the tapes on the graphs made in this lesson.

Record Sheet 4–A

Name: ______________________

Date: ______________________

Matching Our Arms and Our Legs

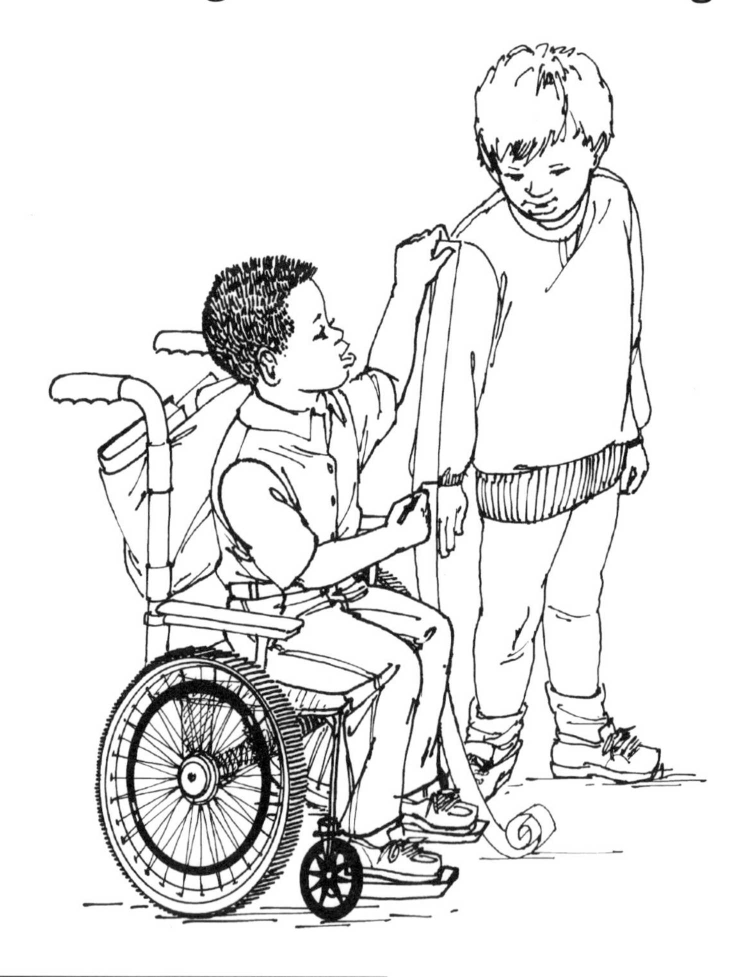

I think ______________________ has the longest arm.

Now match with the tape.

I found out that ______________________ has the longest arm.

Record Sheet 4–A

Name: ______________________

Date: ______________________

Matching Our Arms and Our Legs *(continued)*

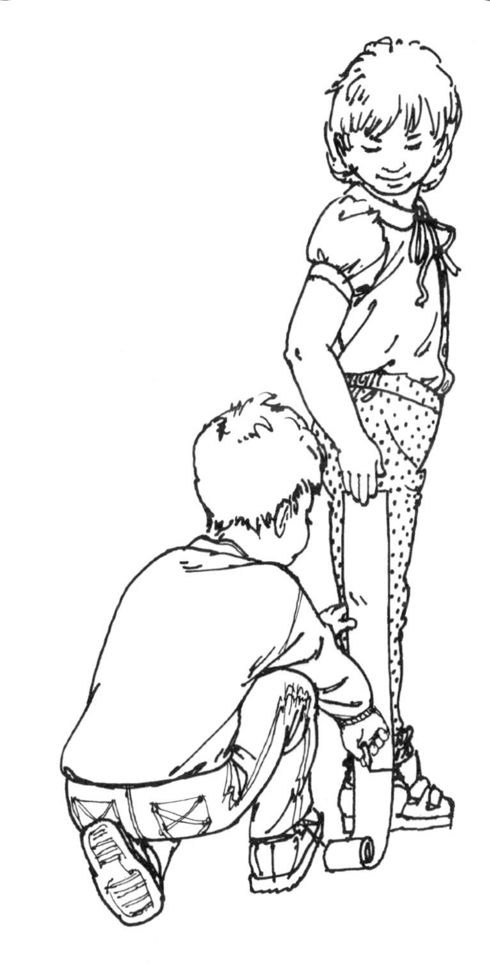

I think ______________________ has the longest leg.

Now match with the tape.

I found out that ______________________ has the longest leg.

Colors
red
yellow
blue
green
purple
orange
brown
black
Today is Thursday,

LESSON 5

Comparing Objects

Overview and Objectives

In the first four lessons, students compared and matched things that were their own size or smaller. In this lesson, students are challenged to match much larger and longer objects. This activity provides an opportunity for students to practice matching beginning and ending points of the objects to the tape and enables you to assess how they apply concepts and skills developed in the first four lessons. As they consider which object might be the longest, students are introduced to the idea of predicting.

- Students discuss the difference between a guess and a prediction.
- Students predict which of several objects is the longest and which is the shortest.
- Students use adding machine tape to match the length of the large objects.
- Students discuss the results recorded on a class graph.

Background

This lesson challenges students to make predictions about the relative size of classroom objects. As they discuss the difference between a guess and a prediction, students may need help in understanding that a prediction is based on previous experiences or observations, while a guess is more random.

As students gain more experience with predicting, you may find that their predictions are closer to their actual results. Some students, however, may continue to make random guesses because of their developmental level. Giving students frequent experience with predicting over time will help them begin to understand the concept.

In this lesson you will want to emphasize that there are different ways to measure the size of objects. For example, you could measure a door from top to bottom or from side to side. Therefore, some students may match the same object but get very different results because they used different beginning and ending points. Let your students know the beginning and ending points for each object they will match with the adding machine tape. This will help them avoid confusion when they construct the graph at the end of this lesson.

Materials

For each student

1 **Record Sheet 5-A: Matching Long Objects**

For every four students

1 roll of adding machine tape
1 pair of scissors
1 marker

For the class

1 sheet of newsprint
1 red marker
2 sheets of paper, 22 × 28 cm (8½ × 11 in) — optional
1 copy of each of the blackline masters **Long Objects**

Preparation

1. Arrange the materials for this lesson in the distribution center.
2. Write "Long Objects" at the top of the chalkboard.
3. Fold the sheet of newsprint lengthwise in half (see Figure 5-1).

Figure 5-1

Folding newsprint lengthwise in half

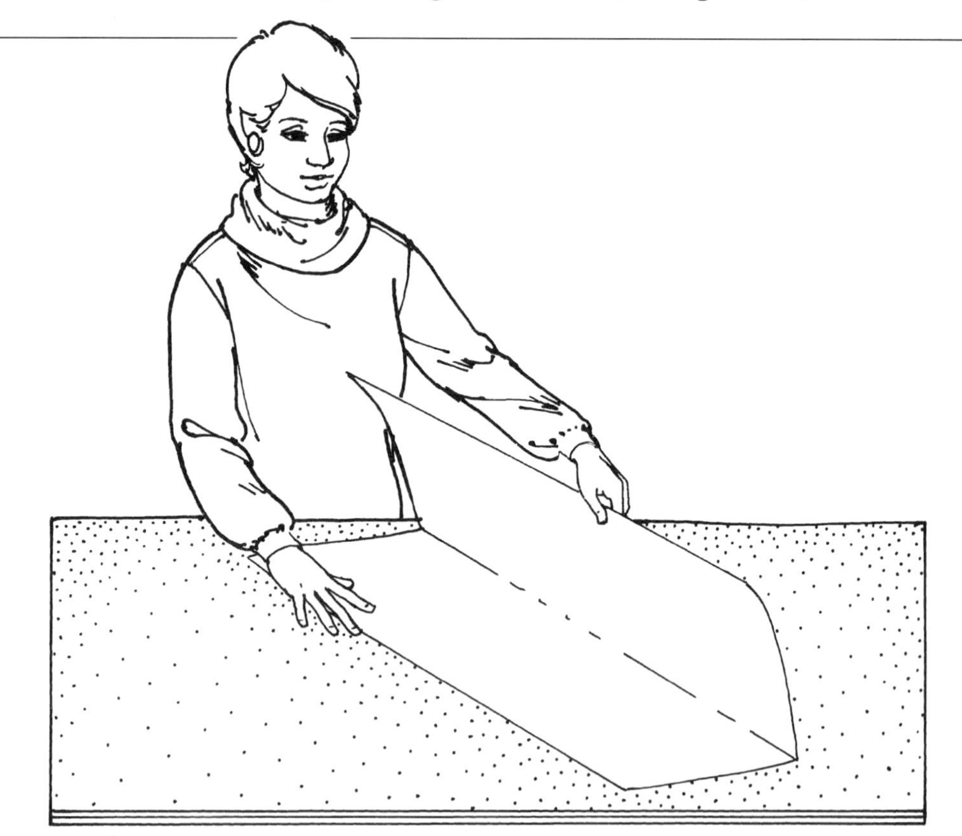

Cut the sheet of newsprint in half. Tape the ends of the pieces together to make one long strip. Orient the strip horizontally and divide it into six equal sections, one section for each group of four. Draw a red horizontal line across it 2.5 cm (1 in) from the bottom. Make a copy of the six blackline masters, **Long Objects** (pgs. 48–53). Glue a long object in each section. Label each section with the name of the object. Title the graph "Comparing Long Objects." Display the graph for the class (see Figure 5-2).

Figure 5-2

Sample "Comparing Long Objects" graph

Note: If you have more than 24 students, you may want to choose two more objects for students to match. Draw each object on a sheet of 22-by-8-cm (8½-by-11-in) paper and attach the two sheets to the end of the graph.

4. Arrange students in groups of four.
5. Copy **Record Sheet 5-A: Matching Long Objects,** on pg. 47, for each student.

Procedure

1. Ask students to think about the lengths they have worked with in previous lessons (arms, legs, and body). Try to focus their ideas on the relative size of the body parts by noting that arms and legs were smaller than their heights. Have students think of things larger or longer than themselves that they could match with the adding machine tape. Record their ideas on the chalkboard. Ask, "How could we find out the length of some of these objects?"
2. Let students know that today they will go on a "matching hunt" and find out the length of some long objects in their surroundings. Ask students to look at the "Comparing Long Objects" graph with the pictures of the objects to be matched. Talk about the fact that the objects can be matched both up and down and side to side. Help your students establish the beginning and ending points for each object.
3. Share with the class an example of a prediction. You look outside in the morning and it is raining. Considering your past experience with rain, what do you think most people will wear to school today?
4. Ask students to talk with their partners about what they think a prediction is. Have partners share their thoughts with the class. Previous student responses have included "It doesn't mean you're always right" and "Things change." Explain that a prediction is a guess based on experience and that today they will be making predictions about the size of long objects. Let students know that all their ideas will be accepted. Reassure them that as they gain more experience with predicting, they may find that their predictions are closer to their actual results.

   Note: You will need to determine how well your students have understood the concept of predicting. If necessary, provide more examples of predicting before going on with the lesson.
5. Distribute and review **Record Sheet 5-A: Matching Long Objects.** You may want to have students find the objects in the classroom to see concrete models before they make their predictions.
6. Give students a few minutes to think about the size of the objects on the graph. Then have students write on Record Sheet 5-A their predictions of which object is the longest and which is the shortest.

7. Assign one of the objects to each group and show them where to find it. Then, share with the class the process they will follow to match the length of their assigned object.
   - With your group, collect your materials from the distribution center.
   - As a group, decide how to determine the length of the object you will match. Think about these questions:
     - What part of the object will you match to determine length?
     - What will be your beginning and ending points?
   - Help your group use the adding machine tape to match the object's length. You may need to remind your group to use beginning and ending points.
   - When you have matched the object's length with the adding machine tape, cut the tape and label it with the object's name.
   - Help your group clean up the materials.
8. As students are working, you may want to circulate to assess the ways they are matching with the adding machine tape.

Final Activities

1. As students finish their matching and recording and return their materials to the distribution center, have them bring their adding machine tape strips with them to the "Comparing Long Objects" graph.
2. Ask students how they used the adding machine tape to determine the lengths of their objects. Ask questions such as the following:
   - How did you determine the length?
   - How does matching the length of something long compare with matching your arms, legs, or body?

   Note: Have students refer to the large object cards glued on the graph to show the beginning and ending points they used when matching.
3. Have students glue or tape their adding machine strip under the appropriate object on the graph. Remind students to attach the strip to the red starting line to help them make fair comparisons.
4. Discuss with students any observations they make about the adding machine strips, such as that a tape is the longest or shortest or that some tapes are the same length.
5. Keep the "Comparing Long Objects" graph posted.

Extension

SOCIAL STUDIES LANGUAGE ARTS

Go to a museum and match the lengths of animal replicas. Have students write a class story abut the animals and illustrate the story with the tapes.

Record Sheet 5–A

Name: ______________________

Date: ______________________

Matching Long Objects

I predict the ______________________ is the shortest object.

I predict the ______________________ is the longest object.

Blackline Master

Long Objects

Blackline Master

Long Objects

Long Objects

Blackline Master

Long Objects

Long Objects

Long Objects

LESSON 6

Matching Distance

Overview and Objectives

In this lesson, which continues to focus on matching, students begin to discover that measuring distance is another facet of measuring length. To do so, they flip toy Flippers™ and use adding machine tape to represent the distance the Flippers™ have traveled. Once again, students apply the concepts and skills they have learned about matching: using a common starting line and beginning and ending points. In the next few lessons, students continue to expand their knowledge of these factors as they investigate ways to measure using nonstandard units.

- Students discuss various ways of matching distance.
- Students explore how to make the Flippers™ move.
- Students compare the distance of three jumps.
- Students use adding machine tape to match distance.
- Students discuss their results and observations.

Background

In past lessons, students have practiced measuring length. Another way to think about length is in terms of distance, or, "how far?" In this lesson, students will think about distance as they measure how far a Flipper™ has traveled.

In the last four lessons, students matched lengths with clear beginning and ending points, such as the wrist and shoulder. In each case, they placed the tape on top of or against the objects they were matching. In this lesson, they discuss that when matching the distance between the starting line and where the Flipper™ lands, there are clear beginning and ending points, but there is nothing concrete between the two points.

Another challenge for students will be to recognize that the length of the tape represents how far the Flipper™ has jumped, or the distance traveled. You can help students make the transition from length to distance by encouraging them to rephrase any comments they make that focus on length. For example, if a student states that his or her tape is longer than another student's, encourage the student to emphasize distance by saying, "My tape is longer," which means "My Flipper™ went farther" or "My tape shows that my Flipper™ went farther."

For information about students' progress, observe students matching and recording their results and review the flipping graphs in this lesson. You may want to refer to the **Assessment** section at the end of Lesson 3 for specific points to note.

Materials

For every two students

1 sheet of newsprint
Glue
1 pair of scissors
2 Flippers™
2 crayons matching the color of the Flipper™ (either red or blue)
2 rolls of adding machine tape
1 resealable plastic bag, 23 × 30 cm (9 × 12 in)

For the class

1 red marker

Preparation

1. Arrange the materials for this lesson in the distribution center.
2. Prepare one sheet of newsprint for every two students. Use a red marker to draw a horizontal starting line 2.5 cm (1 in) from the bottom of each sheet.
3. Arrange students in pairs. Make sure that partners in each pair use a different-colored Flipper™.

Management Tip: You will be asked to group students in the same pairs in Lesson 16, during which the students will also use the Flippers™. You might want to note the names of student groups today in order to make the regrouping simpler at the end of the unit.

Procedure

1. Ask students to think about how frogs move. What words might they use to describe how a frog moves? Does a frog leave any evidence behind to show that it has moved from place to place? Explain that today students will make an object move similarly to the way frogs move.
2. Remind students that while a frog hops from place to place, we most often walk. Discuss with children that when they walk, they usually do not leave a trail unless they have been walking in mud or snow. Ask several students to demonstrate the absence of a trail by having them walk in the classroom from the chalkboard to the door or from the table to the sink.
3. Explain that in today's lesson, students will have an opportunity to determine the distance an object (the Flipper™) has moved. But there won't be any "footprints" left behind for them to follow.
4. Discuss with students the process they will follow (see Figure 6-1).
   - With your partner, collect the materials from the distribution center.
   - Find a place on the classroom floor or on a table to "flip."
   - Write your name below the red starting line on the "flipping graph."
   - Put your Flipper™ on the red starting line and draw a circle around it with your crayon.
   - Take several minutes to practice flipping.
5. After five minutes of practice with the Flippers™, let students know they will now be using their Flippers™ and matching the distances flipped. Share the following process (see Figure 6-2):
   - Place the Flipper™ on the circle you drew on the common starting line.

Figure 6-1

Students flipping

- While your partner observes, flip the Flipper™ and circle the landing spot with your crayon.
- Now switch places and have your partner do the same thing.
- Each of you take two more turns, circling each landing spot. Make sure you use your same beginning circle for each flip.
- After you have taken your three flips, decide which flip was the farthest.
- Using the adding machine tape, match the distance from the circle at the starting line to the circle of your longest flip.
- Cut the adding machine tape and label it with your name.

Final Activities

1. After students have matched the distances of their longest flips, have them find one student with a longer adding machine tape and one student with a shorter tape. Discuss as a class questions such as the following:
   - Why are some tapes longer or shorter than others?
   - What information do the tapes give you about the distance flipped?
   - How did the distance of your three flips compare?
2. Ask students to discuss other ways they could determine the distance their Flippers™ have traveled.
3. Have students use the glue to attach their adding machine strips to their flipping graphs. Save the flipping graphs for use in Lesson 16.
4. Have students place all the materials back in the plastic bags and return the bags to the distribution center.

Figure 6-2

Flipping and matching

Extensions

MATHEMATICS

1. Make a large X out of construction paper and place it on the classroom floor. Make a graph listing different locations in the classroom, such as the door, the reading corner, and the sink. Have children predict which locations will be the closest to and farthest from the X. Next, ask students to work in pairs. Have one student stand on the X and hold the end of the adding machine tape while the other student pulls the roll to the team's designated location. Ask students to cut and label the tapes and attach them to the graph.

LANGUAGE ARTS **ART**

2. Read *How Much Is A Million?* by David M. Schwartz. After reading the book, have students discuss with a partner long distances such as from the earth to the moon or from the East Coast of the United States to the West Coast. Then have students draw pictures illustrating their ideas.

LESSON 7

Using Our Feet to Measure

Overview and Objectives

In the last five lessons, students have compared lengths by matching. In this lesson, students make the transition from matching to measuring by quantifying nonstandard units of measure. In this case, each student's foot is the nonstandard unit. Students take five heel-to-toe steps on a strip of adding machine tape and match the length between the beginning and ending points. By comparing the lengths of the five steps and discussing the results, students begin to recognize that using nonstandard units of measure produces varied results.

- Students take five heel-to-toe steps and match the length covered.
- Students record the results on a graph.
- Students discuss and compare the results.
- Students discuss why the tapes are different lengths.
- Students listen to and discuss a story about nonstandard units of measure.

Background

In today's lesson, students use their feet to measure. While each student uses the same unit of measure—"my foot"—the units are nonstandard because students have different-sized feet. When students measure the same object, the use of the foot as a unit produces varied results.

As students walk five steps heel to toe and match this distance with the strips of adding machine tape, they are introduced to one of the main concepts of measuring: units are placed end to end. If you look at any measuring tool (for example, a ruler, meterstick, or yardstick), this is easy to see. The importance of this fact may not be immediately evident to your students; however, during the next few lessons their awareness of the need to place units end to end will increase.

Students face another challenge in this lesson—figuring out how to count each step that they take on the tape. The process of attempting to qualify a length may produce confusion. For example, some students may not count their first step onto the tape as "one" but will start to count as they take their second step. As they progress through the activity, students will discuss that it is important to start counting with the placement of the first unit, or the first foot.

Materials

For every two students

1 roll of adding machine tape
1 pair of scissors
1 crayon

For the class

2 sheets of newsprint
1 marker
1 roll of masking tape
1 large beach umbrella (optional)
1 copy of "The Very Big Umbrella" (see pgs. 66–68)

Preparation

1. Arrange the materials for this lesson in the distribution center.
2. Create a graph by folding two pieces of the newsprint lengthwise and cutting them in half. Tape the two strips end to end. Label the graph "Comparing Feet." With a red marker, draw a horizontal line 2.5 cm (1 in) from the bottom of the chart paper. Divide the strip into equal sections, one section for each student. Each section should be wider than the width of the adding machine tape. (You may not need to use all of the chart paper.) An example of the completed graph is shown in Figure 7-1.

Figure 7-1

Sample graph

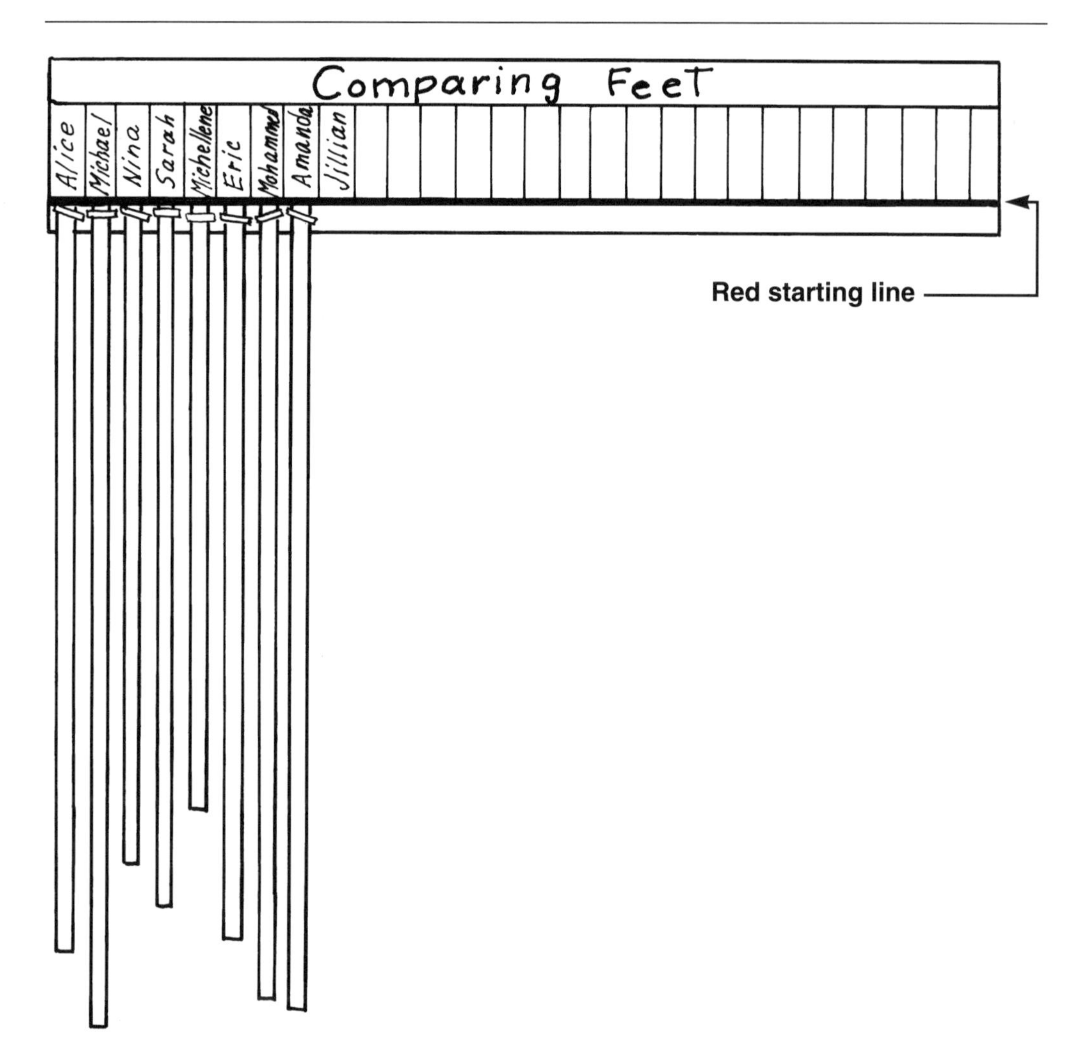

3. Hang the graph in the classroom.
4. Make an X out of masking tape and place it on the floor several feet from the classroom door. Be sure you can track a straight line from the X to the door.
5. Arrange students in pairs.
6. If possible, bring in a beach umbrella to show students before you read them "The Very Big Umbrella," on pgs. 66–68.

   Note: You may want to copy "The Very Big Umbrella" for students so they can follow along as you read. If you are using the Student Notebooks, the reading selection appears on pgs. 6–8.

Procedure

1. Ask the class to help you solve a problem. Explain that you would like to determine the distance from the X on the floor to the classroom door. But you forgot to bring any adding machine tape along. How could you get the information you need? Have students take a few minutes to talk with their partners about possible solutions.
2. Have students share their thoughts with the class. You may want to have students demonstrate some of the methods they have suggested.
3. Let students know that long ago, people did not have measuring tools. One way that people determined length was by using their feet. They would measure by walking slowly, placing the heel of one foot in front of the toes of the other foot and making certain that the heel of one foot always touched the toes of the other one. Each time they picked up a foot and continued the pattern, they would count. This way, they always had a measuring tool wherever they went.
4. Ask several students to demonstrate how to measure using their feet. Make certain that they place their feet close together as they count.

   Note: Some students may be confused about when to start counting and may not count the beginning foot. They should begin counting when they take their first step. You may want to match those students who demonstrate understanding of this concept with those students experiencing difficulty.
5. Let students know that in this lesson, they will work with a partner to "walk and count" a strip of adding machine tape.
6. Discuss with students the process they will follow.
   - With your partner, collect the materials from the distribution center.
   - Have your partner use a small piece of masking tape to attach one end of the tape to the floor. Unroll the adding machine tape on the floor.
   - Stand at the beginning of the adding machine tape. Put your foot on the tape with your heel at the end of the tape, and count one step. Using the "heel-to-toe" method, keep taking steps until you have counted to five (see Figure 7-2).
   - As you take steps, have your partner unroll more tape as needed and help you count.
   - Once you have counted five steps, have your partner use a crayon to mark the ending point on the tape. Then have your partner cut the tape at the ending point.
   - Label the tape with your name.

- Switch places with your partner and repeat the activity.
- Clean up your materials and return them to the distribution center. Bring your strip of tape back to your seat.

Figure 7-2

How to walk and count

7. As the students do the activity, circulate to assess how they are walking and counting on the adding machine tape.
8. Bring the class together. Show students the "Comparing Feet" graph. Point out that each student has one section of the graph. Ask students to review the purpose of the red starting line.
9. Beginning at the red starting line, tape the strips to the graph. Place one strip in each section and label each section with the appropriate student's name.
10. Remind the class that everyone counted the same number of steps—five—and that the strips are all taped to the same starting point. Ask students why they think the strips are different lengths.

Management Tip: You may want to do the **Final Activities** during language arts.

Final Activities

1. Let students know that they will listen to a story about measuring. Show them the beach umbrella you brought in and explain what it is usually used for.
2. Read "The Very Big Umbrella" aloud to students. Invite students to think about and discuss what happened to Marcus and Kate as they were measuring the umbrella and the box. Ask questions such as the following:
   - How did Kate and Marcus measure the umbrella and the box?
   - Why was the umbrella too big for the box?
   - How could Kate and Marcus solve the problem?

   Have students use the beach umbrella to demonstrate their ideas.

Extension

LANGUAGE ARTS MATHEMATICS

Read *How Big Is a Foot?* by Rolf Myller. Place students in groups of four. Have each group select an object to measure using their feet. As each person in the group measures the object, have one person in the group record the results. Ask students to share their results with the class and discuss a way to make the measurements more consistent.

Assessment

As you observe students during the measuring and reading activities, note the following:

- Do students place their heels and toes close together?
- Do they begin counting as they take their first step onto the tape?
- Do they help their partners count each step?
- Did students recognize the reasons why, in the story, the umbrella didn't fit in the box?
- Were students able to identify ways for Kate and Marcus to solve their problem?

Reading Selection

The Very Big Umbrella

It was raining outside and Marcus was looking out the window. He was trying to think of a birthday present for his father. Suddenly, his father came running in the house. He was very wet. "I lost my umbrella and now I'm wet," said Mr. Hill. Marcus said to himself, "That's it! I'll get Poppa a big, new umbrella."

The next day, Marcus went to the store to look for a big umbrella. He found one that was just the right size, but it was too much money. Marcus felt sad. "What will I do now?" he thought. As he walked home, he had an idea. "I will make an umbrella for Poppa."

In the garage, Marcus found an old beach umbrella. It was very big, but it had many holes in it. Marcus put some tape on the holes. He found some paint and painted it. Marcus looked at the umbrella and smiled. "Poppa will really like this umbrella, and it is so big, he won't lose it."

"Now I will have to find a box to put it in," he said. Marcus looked in the house and the garage but all the boxes were too small. Marcus had an idea. "I will call Kate and ask her for a box."

Marcus called Kate on the telephone and told her about the umbrella. "I need a big box to put it in," he said. "How big is the umbrella?" asked Kate. Marcus put the telephone down and ran to the garage. "Hmm," he thought. "How can I measure this?" Marcus put the umbrella on the floor and lay down next to it. The umbrella was longer than Marcus. He ran back to the telephone.

"Kate, the umbrella is longer than me," he said. "But Marcus," said Kate. "I still don't know how big the box should be." Marcus put the telephone down and ran to the garage again. This time he used his feet to measure the big umbrella. He went heel to toe from the top of the umbrella to the end. As he walked, he counted each step. Marcus ran back to the telephone.

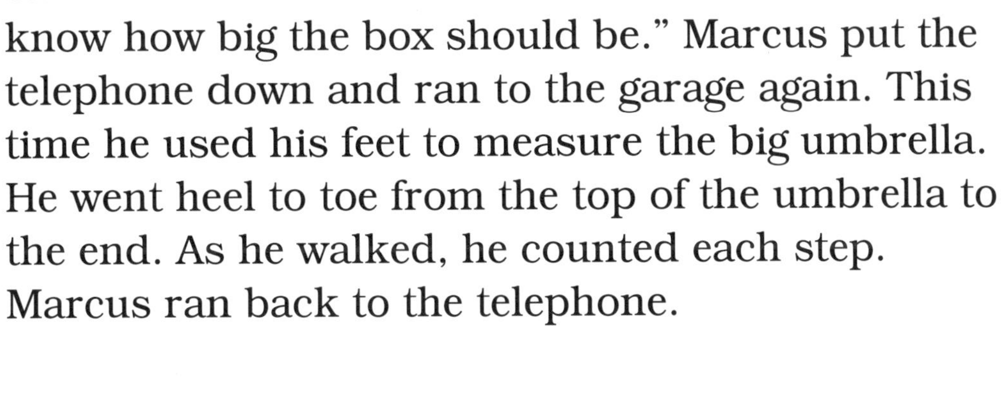

"Kate, this time I measured the umbrella with my feet and counted as I walked," said Marcus. "What did you get?" asked Kate. "Nine of my feet," he said. "Okay," said Kate. "I will look for a box that the umbrella will fit in."

Later, Kate brought the box over and they went into the garage. Marcus and Kate tried to put the umbrella into the box. It did not fit. The box was too small! "Oh, no," said Marcus. "This box is too small. How could that be? I told you it had to be nine of my feet long."
"It is," said Kate. "I measured just like you and counted each step I took until I counted to nine."

Marcus decided to measure the box with his feet. He counted each step as he measured. When he got to the end of the box, he had only counted eight steps. "Kate," said Marcus, "This box is not nine feet—it's only eight feet. The box is too short." Kate decided to measure the box again. She counted as she took each step. When Kate got to the end of the box, she counted nine steps and said, "Marcus, it is nine feet. I measured just the same as you."

a b c d e
f g h i j
k l m n o
p q r s t
u v w x y z

LESSON 8

Using Different Standard Units of Measure

Overview and Objectives

In Lesson 7, students discovered that one reason for varied results was that their unit of measure—the foot—was not a standard size. In Lesson 8, students continue to get varied results as they use different sets of standard units to measure the same object. For example, students who use pencils to measure achieve different results from students who use spools to measure. In addition, students now begin to think about why it is important to label their results with the name of the unit they have used to measure. These experiences lay the foundation for Lesson 9, in which the entire class will use the same standard unit to measure.

- Students discuss what they now know about measuring.
- Students measure the lengths of objects using various sets of standard units.
- Students record the results.
- Students compare and discuss their results.

Background

Students have now developed an awareness of some of the factors that affect the results of a measurement: using beginning and ending points, counting every unit, and placing the units end to end. In this lesson, students use a variety of sets of uniform, or standard, units to measure the lengths of objects. For example, all the coffee stirrers in the lesson are the same length and size. However, as students will observe at the end of the activity, the results will vary when they measure the same object with different standard units. Students begin to understand that even when they apply consistent beginning and ending points, place the units end to end, and use standard units, the results still vary because there is no single class standard.

Materials

For each student

1 pencil
1 **Record Sheet 8-A: Measuring Different Objects**

For the class

2 sheets of newsprint
1 marker
90 wood coffee stirrers
90 unsharpened pencils

90 plastic spoons
90 toothpicks
90 small wood spools
30 resealable plastic bags, 23 × 30 cm (9 × 12 in)

Preparation

1. Arrange the materials for this lesson in the distribution center (see Figure 8-1). Count out each of the five items in bundles of 15 and place them in resealable plastic bags. Each student will take one bag of either the coffee stirrers, pencils, spoons, toothpicks, or spools.

Figure 8-1

Distribution center

2. Label one sheet of newsprint "What Did You Find Out." Hang it in the classroom.
3. Copy **Record Sheet 8-A: Measuring Different Objects,** on pg. 75, for each student.

Procedure

1. Remind students that in Lesson 7, they measured by using their feet. Ask them to share their thoughts on any new information they have gained about measuring. (In the past, students have said, "We need to keep our feet close together when we go heel to toe," "We got different answers because everyone's feet are different sizes," "I started on the beginning of the strip and stopped at the end," and "I can use my feet to measure things.") Record these thoughts on the other sheet of newsprint and save it for assessment purposes.
2. Let students know that today they will be using many different units (coffee stirrers, pencils, spoons, toothpicks, or spools) to measure classroom objects. Ask students to take a few minutes to choose four objects in the classroom they would like to measure. Encourage students to select a variety of objects. Make sure at least two students measure each object.
3. Show students the five measuring units. Let students know that they will choose one of the sets of units and use it to measure their objects. Each student will have 15 of the measuring units he or she has chosen.
4. Distribute **Record Sheet 8-A: Measuring Different Objects.** Show students the spaces to write their predictions and results.
5. Describe the process students will follow.
   - Collect a bag of 15 items from the distribution center.
   - On the record sheet, write the name or draw a picture of the first object you are going to measure.

- Observe the object and predict its length in terms of units.
- Write your prediction on the record sheet.
- Measure the object with the units.
- Write your results on the record sheet.
- Repeat the steps with the next object you will measure.

6. Circulate as the students work and make sure at least two students measure each object.

Final Activities

1. Focus attention on the class chart labeled "What Did You Find Out?" Ask students to share their thoughts about measuring with coffee stirrers, pencils, spoons, toothpicks, and spools. Ask questions such as the following:
   - When you measured the same object as someone else, but with different units, were your results the same?
   - Were your results different? Why?

   Record students' thoughts on the chart (see Figure 8-2).

Figure 8-2

Sample chart

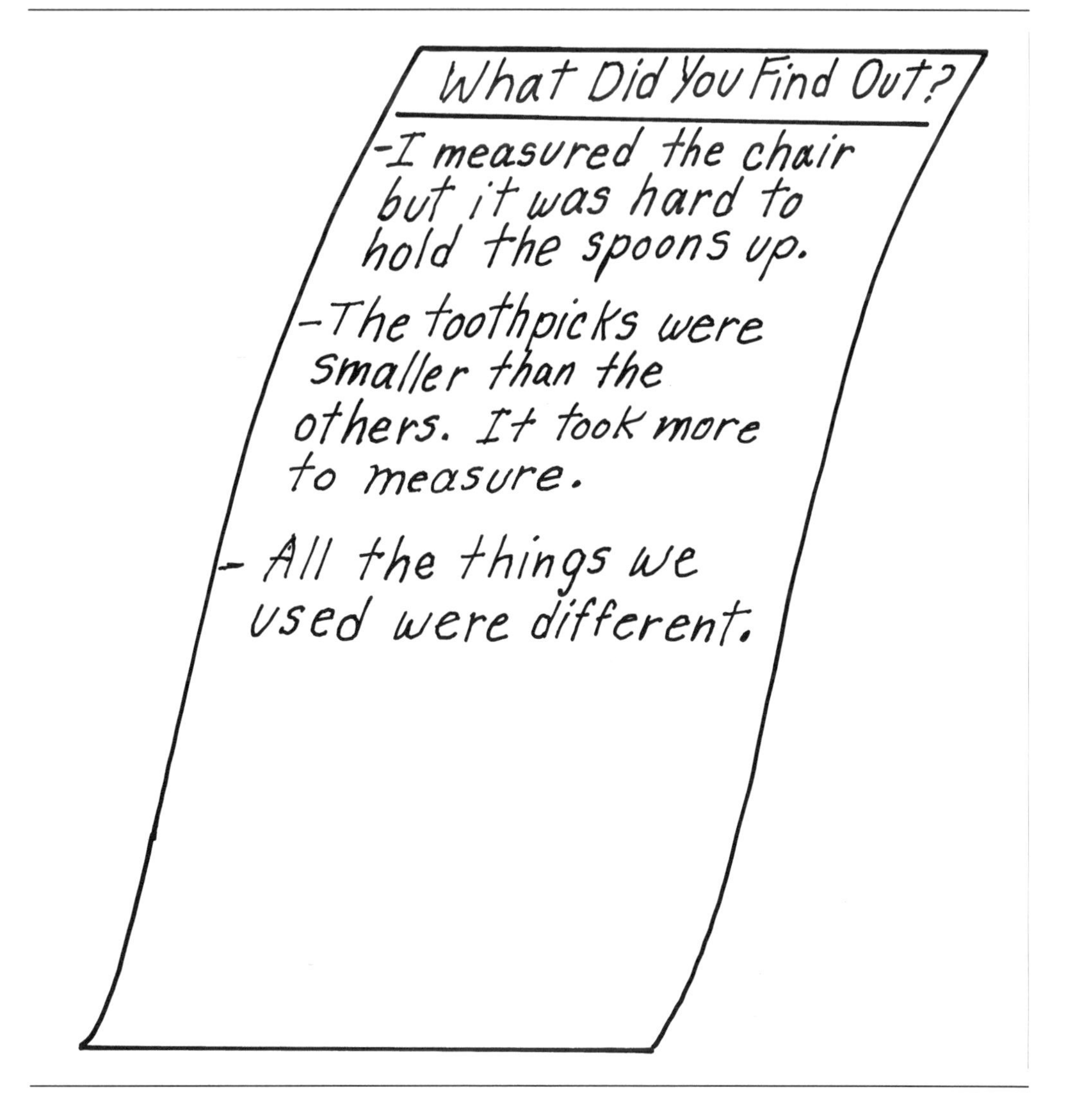

2. Have students look at their record sheets. Then have them exchange sheets with someone close by. Ask questions such as the following:
   - Does the record sheet let you know the length of the objects?
   - Do you know what unit was used?
   - What does knowing the unit enable you to do?
3. Hold up a spoon and a toothpick. Ask students to compare the two units.
4. Next, hold up a pencil and a spool. Ask students to compare the two units.
5. Ask students to focus on the size of one of the sets of standard units (for example, the spools). Ask questions such as the following:
   - Are all the units the same size?
   - When you compared your results with those of another pair of partners who measured the same objects, what did you find out?
   - Were your results the same as theirs? Why do you think that happened?
6. Save the "What Did You Find Out?" chart for use in the next lesson.

Extension

Have students draw themselves measuring in the lesson. Ask them to write a sentence describing what they did.

Record Sheet 8–A

Name: ______________________

Date: ______________________

Measuring Different Objects

The unit I used to measure is a ______________________.

1.	2.
______ prediction ______ measurement	______ prediction ______ measurement
3.	4.
______ prediction ______ measurement	______ prediction ______ measurement

COLOR WORDS
TOWNHOUSE

LESSON 9

Measuring with a Standard Unit

Overview and Objectives

Students now expand their understanding of measuring by using a class standard unit—a coffee stirrer—to quantify the lengths of objects. They begin to recognize that when they use one standard unit, they have a common language with which to compare results. In addition, students continue to think about why it is helpful to label their results with the name of the unit they have used to measure.

- Students use coffee stirrers to measure the lengths of objects.
- Students record their results by gluing the coffee stirrers to a length of adding machine tape matched to the length of the object.
- Students record, compare, and discuss their results.
- Students label their results with the name of the unit.

Background

By using one standard unit to measure, students learn more about the factors that affect measuring. For example, students discuss that since the coffee stirrers are used by everyone and all the stirrers are the same size, the measurements are less divergent than when students were using five different sets of uniform units, as they did in Lesson 8.

This lesson also introduces students to another important concept in measuring: by using a common standard unit of measure, one can communicate and compare results more easily. It is easier, for example, to make meaningful comparisons between numbers of coffee stirrers than between numbers of spoons and numbers of pencils.

Materials

For every two students

15 coffee stirrers
1 roll of adding machine tape
1 bottle of glue
1 resealable plastic bag, 23 × 30 cm (9 × 12 in)

For the class

2 sheets of newsprint
1 red marker
1 sheet of white paper, 22 × 28 cm (8½ × 11 in)
1 Post-it™ notepad, 8 × 13 cm (3 × 5 in)
"What Did You Find Out?" chart (from Lesson 8)

Preparation

1. Arrange the materials for this lesson in the distribution center. You may want to count out the coffee stirrers ahead of time and bundle them together.
2. Prepare a new graph similar to those you used in Lessons 5 and 8. Fold a piece of newsprint in half lengthwise and cut it into two equal sections. Tape the ends together to make one long strip. Use the marker to divide the graph into four equal sections. Draw a red starting line 2.5 cm (1 in) from the bottom of the graph. Title the graph "Measuring with Coffee Stirrers."
3. Number the sections of the graph one through four (see Figure 9-1). Choose four objects in the classroom for the students to measure. Write the name of an object in each section of the graph.

   Note: You can increase the number of sections on the chart to accommodate the size of your class. Each section should have enough room for four or five strips of adding machine tape.

Figure 9-1

Sample graph

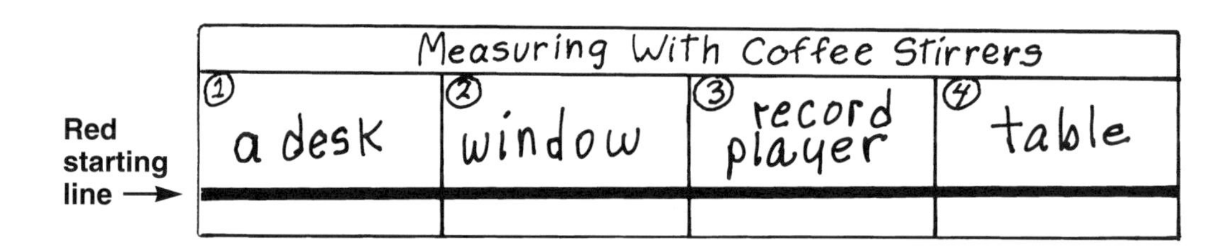

4. Write "1" on a Post-it™ note and attach it to the object you have selected for section one of the "Measuring with Coffee Stirrers" graph. Do the same for the objects numbered 2, 3, and 4.
5. Cut the sheet of white paper into 16 3-by-8-cm (1-by-3-in) strips. Divide the strips into four groups. Write the number "1" on the strips in one group. Continue numbering the strips in groups 2 through 4.
6. Arrange students in pairs.
7. Label the other sheet of newsprint "What We Know about Measuring."

Procedure

1. Review and discuss the "What Did You Find Out" chart from Lesson 8. Remind students that when they measured objects in Lesson 8, they used different sets of units (pencils, toothpicks, coffee stirrers, spools, and spoons). When they compared the measurements, they were very different. Why were the measurements different? What affected the results?
2. Let students know that today they will all use the same set of units—coffee stirrers—to measure objects.
3. Have each pair of students select a numbered strip. Ask students with strip number 1 to measure the length of the object labeled 1. Have students do the same for the objects labeled 2, 3, and 4. To help students make fair comparisons, you may want to help them identify the beginning and ending points of the objects to be measured.
4. Share with students the process they will follow.
   - Collect your materials from the distribution center.
   - Measure your object with the coffee stirrers by placing them end to end and counting.

- Match the length of the line of coffee stirrers with the adding machine tape.
- Cut the adding machine tape and place the coffee stirrers end to end on the tape. Make sure there is no space between the ends of the coffee stirrers.
- Glue the coffee stirrers end to end on the tape.
- Write your name on the tape. Next to your name, write the number of coffee stirrers you used.
- Next to the number, write the words "coffee stirrers."

5. Have students glue their tapes on the "Measuring with Coffee Stirrers" graph. Beginning on the red starting line, students measuring the object labeled 1 will attach their tape in the section with the number 1, and so on (see Figure 9-2).

Figure 9-2

Sample graph with tapes

Final Activities

1. Focus students on the "Measuring with Coffee Stirrers" graph and ask questions such as the following:
   - Which object took the most coffee stirrers to measure?
   - Which object took the fewest coffee stirrers to measure?
   - Which objects took about the same number of coffee stirrers to measure?
2. Focus students on the things they do when measuring that enable them to make fair comparisons. Use questions such at the following:
   - What kinds of things do you do when measuring to help you make fair comparisons?
   - Why do you need to use the same standard unit when measuring?
   - How did writing the words "coffee stirrers" next to the number help you compare your results?
   - How have you used the red starting line?

   Record their ideas on the newsprint labeled "What We Know about Measuring." Save this chart for use in Lesson 10.
3. Save the coffee stirrer strips for use in Lesson 11.

Extension

LANGUAGE ARTS SOCIAL STUDIES

Read *How Big Were the Dinosaurs?* by Bernard Most (see Bibliography). Have students work in groups of four to select and research a type of dinosaur. Students can then use the coffee stirrers to estimate the size of the different dinosaurs.

LESSON 10

Exploring with Unifix Cubes™

Overview and Objectives

In this lesson, students are introduced to a measuring unit that they will use in the next six lessons: Unifix Cubes™. By comparing the use of the coffee stirrers with the use of the Unifix Cubes™, students may identify the advantages of units that can be connected and stacked. In the next lesson, students are introduced to a strategy for counting large numbers of Unifix Cubes™ when using them to measure longer objects.

- Students reflect on what they have learned about measuring in the last three lessons.
- Students compare coffee stirrers with Unifix Cubes™ as units of measure.
- Students measure using Unifix Cubes™.
- Students discuss their measuring experiences.

Background

In the last few lessons, students have developed a growing understanding of the factors that affect the result of a measurement. In Lesson 9, they explored using coffee stirrers—a standard unit of measure. As they will discuss today, however, the coffee stirrers have certain disadvantages. For example, it can be very difficult to measure height using the coffee stirrers, which are not designed to connect or stack. On the other hand, Unifix Cubes™, the unit of measure that they use in this lesson, can be connected and stacked, making it easier to measure both vertically and horizontally more precisely.

Every student will have 20 Unifix Cubes™, and you may observe some students counting every cube as they measure. You may also observe some students organizing the cubes into sticks of 10 and counting by tens. The next lesson will formally introduce counting by tens as a strategy for measuring and for counting large numbers of Unifix Cubes™.

Materials

For each student

20 Unifix Cubes™ (10 each of two colors)
1 resealable plastic bag, 23 × 30 cm (9 × 12 in)

For the class

1 "What We Know about Measuring" chart (from Lesson 9)

Preparation

1. Select two colors of Unifix Cubes™. Arrange them in the distribution center, with two colors in each bag.
2. Arrange students in pairs.
3. Display the "What We Know about Measuring" chart from Lesson 9.

Procedure

1. Focus students on the "What We Know about Measuring" chart. Ask, "What did you do when you were measuring with the coffee stirrers that enabled you to share your results and make fair comparisons?" Invite students to share their ideas. Student responses may include, "We placed our coffee stirrers end to end," "We used the same beginning and ending points as everyone else," and "Our coffee stirrers were all the same size."
2. Explain that today students will explore another unit of measure. Show them a bag of Unifix Cubes™. Challenge students to compare the characteristics of the coffee stirrers with the characteristics of the Unifix Cubes™. Ask questions such as the following:
   - Can you put the coffee stirrers end to end? Can you put the Unifix Cubes™ end to end?
   - Are the coffee stirrers all the same size? Are the Unifix Cubes™ all the same size?
   - Can the coffee stirrers be hooked together? Can the Unifix Cubes™ be hooked together?
3. Ask students which unit would be easier to use to measure the height of a door—the coffee stirrers or the Unifix Cube™. (If students suggest the coffee stirrers, you may want to also let them use coffee stirrers to measure the classroom objects in this lesson.)
4. Help students recognize that one advantage of Unifix Cubes™ is that students can connect and stack them. This allows them to measure the height of objects, such as doors, that cannot easily be placed lengthwise on the floor.
5. Have students collect their materials from the distribution center.
6. Give students some time to use the Unifix Cubes™ to measure classroom objects (see Figure 10-1).

Final Activities

1. Ask students to share how they have used the Unifix Cubes™ to measure. (Some students have chosen to match only those objects that were the same length as the 20 Unifix Cubes™. Others counted by tens or broke the Unifix Cube™ sticks apart and counted them as single units.)
2. Now invite students to describe how using the cubes is different from using the coffee stirrers. Previous student responses have included, "The Unifix Cubes™ were easier to connect," "I could stand the cubes up to measure the wall," and "The coffee stirrers were very thin and hard to stack."
3. Have students collect materials and return them to the distribution center.

Figure 10-1

Measuring with Unifix Cubes™

Extensions

MATHEMATICS

1. Have students bring in items that are stackable, such as boxes, cups, or cans. Students can use these items to measure objects vertically and horizontally. Then, have students create a chart showing the different units and the objects they have measured.

MATHEMATICS

2. Set up a Unifix Cubes™ measuring center. Fill tubs with Unifix Cubes™. Have the students use the cubes to measure objects in the classroom.

MATHEMATICS

3. Have students use the coffee stirrers to measure their heights. Glue the coffee stirrers to strips of adding machine tape. Then ask students to use the Unifix Cubes™ to measure their heights. Have them compare the strips with the cubes.

BUGLE

LESSON 11

Counting Large Numbers of Units

Overview and Objectives

In this lesson, students apply some of Lesson 10's discoveries about using Unifix Cubes™ to measure vertically and horizontally. Now they will measure longer objects and they will have more Unifix Cubes™ to count. To simplify the process of counting so many cubes, students are introduced to the idea of grouping the cubes into tens.

- Students measure long objects using Unifix Cubes™.
- Students record their results.
- Students discuss why it is useful to group cubes into tens when measuring.

Background

In measuring longer objects and, therefore, using more measuring units to measure, students are challenged to simplify the process of counting the Unifix Cubes™. Today, they are introduced to the strategy of grouping the Unifix Cubes™ into tens and counting by tens.

During the discussion at the end of this lesson, you will probably observe students recording their measurements in a variety of ways. In fact, students may be confused about how to deal with their measurements. For example, when measuring an object that is 73 Unifix Cubes™ long, students may be able to count by tens to 70, but then not know how to account for the three additional cubes. Take your students' own experiences and developmental levels into account when guiding the class. For example, some teachers encourage students to record their results in the terms "70 cubes and 3 loose cubes" or simply as "73."

If your students have not had much experience counting by tens, consult the **Extensions** on pgs. 89–90 for suggested activities to help them practice this skill.

Materials

For each student

1 **Record Sheet 11-A: Predicting and Measuring with Unifix Cubes™**
1 pencil
Crayons

For every two students

100 Unifix Cubes™ (50 each of two different colors)
1 resealable plastic bag, 23 × 30 cm (9 × 12 in)

For the class

Coffee stirrer strips (from Lesson 9)

Preparation

1. Arrange the Unifix Cubes™ in the distribution center. Every two students will use 100 Unifix Cubes™, 50 each of two different colors. Connect the cubes into sticks of 10. Place 10 sticks in a plastic bag for each pair of students.
2. Arrange students in pairs.
3. Copy **Record Sheet 11-A: Predicting and Measuring with Unifix Cubes™** for each student.

Procedure

1. Review Lesson 10 with students by asking what made the Unifix Cubes™ easy to use for measuring. Remind students that one advantage is that the cubes can be connected. Have students think of the number of cubes that could be connected in a group to make counting easy. (You may want to remind students of prior experiences in your classroom with groups of 10). Have students share their thoughts with the class.
2. After showing students a stick of 10 Unifix Cubes™, ask, "How could you use this to measure objects?" Write their ideas on the chalkboard. Let students know that in this lesson they will use 100 Unifix Cubes™, connected in sticks of 10, to measure objects in the classroom.
3. Distribute **Record Sheet 11-A: Predicting and Measuring with Unifix Cubes™**. Point out that it has four boxes, each with a space to write a prediction and a result.
4. Let students know the process they will follow (see Figure 11-1).
   - Collect your materials from the distribution center.
   - With your partner, decide on four long objects in the classroom you would like to measure.
   - On the record sheet in box 1, write your prediction of the length of the first object.
   - Measure the object with the Unifix Cubes™ in sticks of 10.
   - Record your results in box 1.
   - Write the name of the object or draw its picture in the box.
   - Do the same for the other three objects.
5. Circulate as the students are predicting and measuring with the Unifix Cubes™. Help any students who are confused about how to use the Unifix Cube™ sticks to measure or how to record the measurements. For assessment purposes, you may want to record in a log your observations about specific students' progress in measuring.

Final Activities

1. Have students talk with their partners about how they used the Unifix Cube™ sticks to measure. Have them discuss questions such as the following:
   - What was the longest object measured?
   - What was the shortest object measured?
   - How did you count the Unifix Cubes™ to get your results?
2. Bring students together as a class to discuss these questions.

Figure 11-1

Measuring with more than 10 cubes

3. Have students use the Unifix Cubes™ to measure their coffee stirrer strips from Lesson 9. Have students compare the results. Ask questions such as the following:
   - How is measuring with the Unifix Cubes ™ different from measuring with the coffee stirrers?
   - Which was easier to use? Why?

Extensions

MATHEMATICS

1. Have students use the Unifix Cube™ sticks to measure something larger than themselves. For example, have them measure the hallway, a lunchroom table, or the length of a school bus. Discuss any problems the students had with measuring. How did they try to solve the problems?

MATHEMATICS

2. Help your students practice counting by tens by doing the following activities:
   - Make a counting chart from 0 to 100 and circle the numbers 10, 20, . . . 90, 100 in red. Have students use the chart to practice counting by tens.
   - Have students count 10 items (for example, beans, corks, marbles) and place them in a cup. Have students continue until they have counted 10 cups of 10 items.

- Make flash cards with the numbers 10, 20, . . . 90, 100 written on one side. On the other side of each card, draw a picture of the number of cups needed to match the written number. Then have each student pick a card, look at the written number, and draw that number of cups in a group. Students can check their responses by looking at the pictures on the backs of the cards.

LANGUAGE ARTS

3. Have students write and illustrate a book about objects they have measured and found to be 10 cubes long. It can be called "Everything is 10 Cubes Long."

Assessment

As you observe students in this lesson, think about the following questions:

- Do students use beginning and ending points?
- Do they demonstrate understanding of making predictions?
- Do they count individual Unifix Cubes™, or use the sticks and count by tens?

Record Sheet 11–A

Name: ______________________________

Date: ______________________________

Predicting and Measuring with Unifix Cubes™

1.	2.
________ prediction ________ measurement	________ prediction ________ measurement
3.	4.
________ prediction ________ measurement	________ prediction ________ measurement

car
piano
Ms. Patterson
conference table
soda machine
copying machine
lunch table

LESSON 12

Measuring the Height of the Teacher

Overview and Objectives

In this lesson, students apply what they have learned about using the Unifix Cubes™ in groups of 10 to measure the teacher's height. As students measure a predetermined height, you can assess their ability to make predictions, use beginning and ending points, count by tens, and record results.

- Students predict the height of the teacher in Unifix Cubes™.
- Students use Unifix Cubes™ to measure a strip of adding machine tape that represents the teacher's height.
- Students record their results.
- Students compare and discuss their results.

Background

In today's lesson, each pair of students will probably obtain a slightly different measurement for your height. Keep in mind that as students gain practice and experience with measuring objects, their measurements will be more accurate and more consistent.

Materials

For every two students

1 strip of adding machine tape matched to the teacher's height
100 Unifix Cubes™ (50 each of two different colors)
1 resealable plastic bag, 23 × 30 cm (9 × 12 in)
Crayons

For the class

1 sheet of newsprint
1 marker

Preparation

1. Place the materials for this lesson in the distribution center. As in Lesson 11, every two students will use 100 Unifix Cubes™, 50 each of two different colors. Connect the cubes in sticks of 10. Place 10 sticks in a bag for each pair of students.
2. Have someone help you match a strip of adding machine tape to your height. You will need to cut one strip of tape of this length for each pair of students.
3. Arrange students in pairs.

Procedure

1. Have students review with their partners, and then share with the class, the process they used when they matched the adding machine tape to their heights in Lesson 3.
2. Ask students to describe how they were able to tell which student was the tallest and the shortest.
3. Now ask the students questions such as the following:
   - What have you done that enabled you to make fair comparisons? How was using the Unifix Cubes™ or coffee stirrers different from using the other units?
   - How did you report the length of an object when you used the adding machine tape?
   - How did you report the length of an object when you used the Unifix Cubes™ and the coffee stirrers?
4. Show students one of the tapes you have made and have them guess what object the length of the tape might represent. Then explain that it matches your own height. Share with students that today they will use 10 Unifix Cube™ sticks to measure the height of the teacher.
5. Let students know the process they will follow (see Figure 12-1).
   - Collect your materials from the distribution center.
   - With your partner, write your names on the strip of adding machine tape.
   - Look at the strip of adding machine tape and predict the number of Unifix Cubes™ it will take to match the teacher's height.
   - Write your prediction on the strip next to your name. Remember to label your prediction.
   - Working together, use the Unifix Cube™ sticks to measure the length of adding machine tape.
   - Record your result on the strip of adding machine tape.

Figure 12-1

Using cubes to measure tape

Final Activities

1. After the students measure the teacher's height strip, have partners compare their results with the results of another pair of students.
2. After a few minutes, record the students' measurements on the newsprint.
3. Have students compare their strips by laying them on the floor side by side. Ask questions such as the following:
   - How are all of the tapes the same?
   - Are everyone's measurements the same?
   - If not, why not?
   - How could you check your measurements?

   Note: Some students may not be convinced that the tape actually matches your height. You can hold up one of the height tapes (see Figure 12-2). Or, if you feel comfortable doing so, you can lie on the floor and let students match you with the tape to see for themselves.

Figure 12-2

Matching the teacher's height

Extensions

MATHEMATICS LANGUAGE ARTS

1. Have students compare the teacher's height tape with their own. Ask them to write observations such as "I am taller than Enrique but shorter than Mr. Phillips." Ask students to illustrate their writing.

MATHEMATICS SOCIAL STUDIES

2. Students can match and measure the heights of different people in the school and make an information graph. Comparing the heights of younger students with those of older students and adults will help students recognize the diversity that exists among people.

SCIENCE MATHEMATICS

3. Have students plant fast-growing seeds, such as lima beans, and use Unifix Cubes™ to measure and record the plants' growth.

Assessment

As students apply what they have learned about using the Unifix Cubes™ to measure, you have an opportunity to assess their knowledge and skills. As students compare their measurements, you also have an opportunity to assess their knowledge of the factors that affect the accuracy of a measurement. When you observe students in this lesson, note the following:

- Do students place the Unifix Cubes™ end to end?
- Do students use beginning and ending points?
- Do students count the Unifix Cubes™ by ones or tens?
- What types of observations do students make about the tapes?
- When students count the Unifix Cubes™, what do they do when they come to the end of the tape?

car
piano
Dr. Woody's desk
Mrs. Patterson
conference table
soda machine
Lauren

LESSON 13

Making a Measuring Strip

Overview and Objectives

Students have now begun to recognize the value of using a standard unit of measure, placing units end to end, using beginning and ending points, and counting by tens when measuring. In this lesson, students make a measuring tool that represents 10 Unifix Cubes™. Students then share their ideas on why this tool is more versatile than the Unifix Cubes™. The discussion focuses students on the uses and benefits of this portable measuring tool, which they will use again in the next lesson.

- Students discuss the relative usefulness of the different measuring units they have explored.
- Students use Unifix Cubes™ to make their own measuring strips.
- Students compare their measuring strip with a stick of 10 Unifix Cubes™.
- Students discuss the advantages of using the measuring strips rather than Unifix Cubes™.

Background

Measuring tools were developed as a way to simplify measuring by representing the length of an object that was itself being used as a unit of measure. In Lesson 7, the object being used as the unit of measure was the human foot. Students found that they had to keep picking up and putting one foot after the other when measuring a length longer than one foot. Measuring tools often eliminate this need for iteration. Sometimes, however, iteration is necessary; for example, measuring an object about one yard long with a 12-in ruler. In this case, it would be easier to measure with a yardstick or measuring tape. When we think about measuring, it is important to consider the appropriateness of the measuring tool used.

Measuring tools are used to represent units of measure in both the metric and English systems. Yardsticks and metersticks are both measuring tools that represent certain units—inches and centimeters, respectively—that are placed end to end. In this lesson, your students use Unifix Cubes™ as the class unit of measure to make their own measuring strips. They also discuss how these measuring strips eliminate the need for actual Unifix Cubes™.

Materials

For each student

10 Unifix Cubes™ (five each of two different colors)
2 crayons to match Unifix Cube™ colors
1 measuring strip (blackline master on pg. 102)
1 pair of scissors
1 resealable plastic bag, 23 × 30 cm (9 × 12 in)

For the class

1 marker

Preparation

1. Arrange the materials for this lesson in the distribution center. To save time, you may want to collect the materials for each student and place them in separate containers.
2. Make copies of the blackline master of the measuring strips. Cut out one strip for each student.

Procedure

1. Have students take a few minutes to review the measuring units they have used in past lessons. List the units on the chalkboard. To stimulate discussion, ask students questions such as the following:
   - What did you do with the units to find out the length of objects?
   - What made the units easy to use? Difficult to use?
   - Was one unit better than another? Why?
   - Which unit was the best for you to use? Why?

   Discuss students' responses. Ask students to share some of the advantages of Unifix Cubes™.
2. Hold up a Unifix Cube™ stick. Then hold up a paper measuring strip that is 10 cubes long. Share with students that a measuring strip is a tool composed of units. Ask students how the stick and the strip are alike. How are they different? (Students may say, "They are both 10 cubes long," "You can take the cubes apart, but not the strip," and "Both can be used to measure.") Then ask students to discuss which would fit better in their pockets—the 10 cubes or the paper strip? Help students understand the advantages of the measuring strip—it is portable and flexible.
3. Let students know the process they will follow to make their own paper measuring strip.
   - Collect the materials from the distribution center.
   - Next, make a simple, two-color pattern using the 10 Unifix Cubes™ (for example, alternating red and blue).
   - Use the crayons and copy the pattern onto the paper strip. Label it with your name.

   Note: The two-color pattern makes it easier for students to count the individual cubes.

Final Activities

1. After students have finished, let them use the strip to measure several objects.
2. Circulate as students are using the strip to measure. You may want to record in an observation log how students are using the strip.
3. Ask students to discuss why the strip is easier to use than the Unifix Cubes™.
4. Collect the measuring strips for use in Lesson 14.

Extension

LANGUAGE ARTS MATHEMATICS

Bring in fruits and vegetables such as bananas, carrots, ears of corn, heads of cabbage, or leeks. Read *The Biggest,* by Nicole Irving and John Shadell (see Bibliography). Then have students work with a partner to measure the fruits and vegetables with their measuring strips. Students can draw a picture of each fruit or vegetable and write its measurements.

Blackline Master

Measuring Strips

LESSON 14

Measuring with a Measuring Strip

Overview and Objectives

In this lesson, students are challenged to determine the lengths of objects that are longer than their measuring strips. This problem-solving activity requires them to develop new strategies for measuring objects. The challenge of measuring longer objects sets the stage for Lesson 15, in which students make a measuring tape that is composed of 10 measuring strips, or 100 units.

- Students measure objects that are longer than their measuring strips.
- Students compare and discuss the strategies they used to measure the objects.
- Students discuss the advantages and disadvantages of measuring strips.

Background

In this lesson, you will probably observe students using a variety of strategies for measuring objects longer than the measuring strip. Using the strip accurately to measure objects longer than 10 Unifix Cubes™ requires the process of iteration. Iteration means moving a unit of measure repeatedly while marking beginning and ending points. In the context of this lesson, iteration is the process of marking the spots at the beginning and ending points of the measuring strip, picking the strip up, placing the beginning point of the strip next to the previous ending mark, and repeating this process until the entire length of the object has been measured.

Iteration is a measuring skill that is learned through experience. Although your students have had experience placing numerous units end to end to measure objects, they have not yet measured an object using only one unit—in this case, one measuring strip. As you observe students during the activity, you can expect some students to demonstrate their understanding of the concept of iteration. However, many students will not yet have developed an awareness of this process. These students may flip the measuring strip end over end, randomly slide the strip from one end of the object to the other, or group together several of their collective measuring strips, placing each strip end to end to measure the objects.

Do not expect all your students to apply the process of iteration immediately. This skill will develop as they continue to practice measuring.

Materials

For each student

1 measuring strip (from Lesson 13)
1 **Record Sheet 14-A: Measuring Objects Longer than One Measuring Strip**
1 pencil
Crayons

For the class

1 sheet of newsprint
1 marker

Preparation

1. Arrange students in pairs.
2. Label the sheet of newsprint "Things that Are Long" and hang it in the classroom.
3. Copy **Record Sheet 14-A: Measuring Objects Longer than One Measuring Strip** for each student.

Procedure

1. Let students know that in this lesson they will use the measuring strips from Lesson 13. Have students think of objects to measure that are longer than the measuring strip. Give them a few minutes to discuss their ideas with their partners.
2. Ask students to share their ideas with the class. Record the objects on the "Things That Are Long" chart. Ask students to identify those objects on the chart they could measure today. Circle these objects.
3. Distribute **Record Sheet 14-A: Measuring Objects Longer than One Measuring Strip.**
4. Share with students the process they will follow to measure four of the objects circled on the chart.
   - With your partner, collect the materials from the distribution center.
   - Find an object to measure and predict how many units long it is. Record the prediction on the record sheet in box 1.
   - Now use your measuring strip to measure the object and write the measurement on the record sheet.
   - Next, write the name of the object or draw a picture of it in its box on the record sheet.
   - Fill in your record sheet as you follow this process for three of the other objects circled on the class chart.

   Note: Some students may say an object is five measuring strips long; others may say it is 50 cubes long. Both responses would be accurate.
5. As you circulate around the classroom, encourage students to explore objects of different lengths (see Figure 14-1).

Figure 14-1

Using the measuring strips

Final Activities

1. Ask students questions such as the following:
   - How was using the measuring strip the same as using some of the other measuring units. How was it different?
   - How did you measure objects longer than your measuring strip?

   Choose students to demonstrate the different methods they used to measure.
2. Ask students to share one way the measuring strip was easy to use and one way it was difficult. Some student responses have included, "It was easy because I could measure small things," "It was easy to use because it was just one thing to carry," and "It was hard measuring the chalkboard because the strip was so little."
3. Save the "Things That Are Long" chart for use in the next lesson.

Extensions

LANGUAGE ARTS

1. Read a book that illustrates measuring, such as *Inch by Inch*, by Leo Lionni (see Bibliography).

2. Make a little book for each student. Have students label the book "Long Things." Let students know they will take their book and measuring strip home. There, they can find things that are longer than one strip, measure them, and record the measurements in the book. Then, students can draw a picture of each object. Have them share their books with the class.

MATHEMATICS SOCIAL STUDIES

3. Have your students visit a grocery store, pet shop, or sports store to measure long objects with their measuring strips.

Assessment

As students are measuring, note the following:

- Do they try to match objects the same length as the measuring strip?
- Do the students measure only things smaller than the strip?
- Do they measure objects longer than the strip?
- Are students using iteration?
- When they measure objects longer than the strip, do they place the strip end to end, or do they slide the strip along?
- Do students count by ones, tens, or the number of strips?

Record Sheet 14–A

Name: ______________________

Date: ______________________

Measuring Objects Longer than One Measuring Strip

1.	2.
______ cubes (prediction) ______ cubes (measurement)	______ cubes (prediction) ______ cubes (measurement)
3.	4.
______ cubes (prediction) ______ cubes (measurement)	______ cubes (prediction) ______ cubes (measurement)

car
piano
Dr. Woody's desk
Mrs. Patterson
conference table
soda

LESSON 15

Making a Measuring Tape

Overview and Objectives

In Lesson 11, students discovered that when they measured longer objects, it was easier to use 100 Unifix Cubes™ than to use 10 Unifix Cubes™. In this lesson, they apply that discovery to make a longer measuring tool—a measuring tape composed of 100 units. They also listen to a reading selection about making a long measuring tool. These activities reinforce the concept that using a long measuring tool—the measuring tape—is sometimes easier than using a short measuring tool composed of only 10 units. In addition, students begin to see that a measuring tape has the advantage of being able to measure the outer surface, or periphery, of objects. In Lesson 16, they will revisit the Flipper™ and will have an opportunity to use their new measuring tape.

- Students make a measuring tape.
- Students use the measuring tape to measure objects.
- Students discuss the advantages of using a 100-unit measuring tape rather than a 10-unit measuring strip when measuring longer objects.
- Students measure around objects.
- Students listen to and discuss a reading selection about a long measuring tool.

Background

The measuring tape that students make in this lesson is similar to a measuring tape composed of standardized units, such as centimeters or inches. Both are composed of small units of measure placed end to end. Such tools make measuring long objects easier and reduce the need for iteration. Additionally, as your students may discover in this lesson, the measuring tape enables them to measure the outer surface, or periphery, of objects. You may want to encourage students to measure around objects such as the trash can or the aquarium.

During the activity, your students may describe measuring tapes that they have seen a family member use to measure such things as fabric or lumber. Encouraging students to share these types of observations can help the class recognize the similarities between the measuring tools they have made and the measuring tools in the world around them.

Materials

For each student

10 measuring strips (blackline master)
1 resealable plastic bag, 23 × 30 cm (9 × 12 in)
1 roll of adding machine tape
1 pair of scissors
Red and blue crayons
Glue

For the class

1 "Things That Are Long" chart (from Lesson 14)

Preparation

1. Arrange materials for this lesson in the distribution center. To save time, you may want to collect the materials for each student and place them in separate containers.
2. Display the "Things That Are Long" chart from Lesson 14.
3. Make enough copies of the blackline master on pg. 120 for each student to have 10 measuring strips.
4. Cut the strips on the cutting lines and arrange them in groups of 10.

Procedure

1. Ask students to take a few minutes to focus on the "Things That Are Long" chart from Lesson 14.
2. Remind students that when they used their measuring strips, they measured some things that were longer than each strip. Ask them to describe how they measured these long things. They may wish to demonstrate with their measuring strips.
3. Have students think about measuring with the Unifix Cubes™. Let students discuss the process they used when measuring something longer than one stick of 10 cubes. (Some students have noted that they connected several sticks of 10 cubes and matched them to the object.)
4. Now ask students to discuss what they could do if they were using their measuring strips and needed to measure an object longer than one strip. Help students understand that they can connect the measuring strips together into a measuring tape just as they did with the Unifix Cube™ sticks. Ask students to share some real-life examples of when they have seen a long measuring tape used.
5. Let students know the procedure they will follow to make a measuring tape (see Figure 15-1).
   - Collect your materials from the distribution center.
   - Color 5 strips red and 5 strips blue.
   - Cut out the 10 measuring strips.
   - Place the 10 measuring strips end to end. Alternate red and blue strips.
   - Cut the adding machine tape to match the length of the 10 measuring strips end to end.

Figure 15-1

Making a measuring tape

- Place the measuring strips on top of the adding machine strip, end to end. Make certain that the measuring strips are close together.
- Glue the measuring strips to the adding machine tape.
- Write your name on the measuring tape.

6. As students finish, invite them to use the measuring tape to measure objects in the classroom. Encourage them to try measuring around objects such as the trash can, aquarium, or globe (see Figure 15-2).
7. Have students share their ideas about using the measuring tape. Encourage discussion by asking such questions as the following:
   - How was using the measuring tape different from using the measuring strip?
   - How was it the same?
   - Did you discover anything you could do with the measuring tape that you could not do with the measuring strip?

Figure 15-2

Measuring around things

Final Activities

1. Read "The Long, Long Measuring Tool" to the class.
2. Have students roll up their measuring tapes and save them for use in Lesson 16.

Extensions

MATHEMATICS

1. Have students work in groups of four, eight, or as a whole class to measure very long objects at school. For example, students can use their measuring tapes to measure the hallway, a school bus, or the lunchroom.

LANGUAGE ARTS

2. After listening to the reading selection in this lesson, students can complete the sentence starter, "Next, Keisha and Rachel decided to use their long, long measuring tool to measure ________________________." Have students write a few sentences and draw pictures to accompany them.

Reading Selection

The Long, Long Measuring Tool

Rachel was in first grade and was learning all about measuring. In school, she made a measuring tool. It was made from paper. She put black lines on the tool to divide it into smaller units. Rachel used the tool to measure things such as her cat, her Grandpa's foot, and her mother's long hair. She could measure the height of objects as well as how big they were around.

One day Rachel and her friend Keisha were measuring things in the backyard. "Hey Rachel," said Keisha. "Let's find out how long the slide is on your swing set."

Pause for a few minutes and ask your students the following question:

What do Rachel and Keisha need to remember to do when they are measuring the slide?

Have students share their ideas with the class.

Rachel and Keisha used the top of the slide as the starting point. They used both of their measuring tools and placed them end to end on the slide until they got to the bottom. "Where should we stop measuring?" asked Rachel.

"Let's stop right at this black line. It's at the bottom of the slide and will work well as the ending spot," said Keisha.

When Rachel and Keisha finished measuring, they found that the slide was 15 strips long. "Wow! That's the longest thing we've measured," said Keisha. "Let's find something even longer to measure.

How about the garden hose?" Rachel said that it would take too long to measure the hose even with two measuring strips. "We need to think of an easier way to measure long things," said Rachel. They sat and thought about how they could solve their problem.

Pause for a few minutes and ask students the following questions:

How could you help Rachel and Keisha solve the problem?
What would you use? How would you use it?

Grandpa came outside. "I saw you both sitting here," he said. "You looked so tired and thirsty that I brought you some juice. What have you been doing to make you so tired?" "Keisha and I have been measuring things with our measuring tools," said Rachel. "We've been measuring long things like the slide. Now we want to measure the garden hose, but it will take a long time."

Grandpa, Rachel, and Keisha sat on the steps and thought about how to solve the problem. Suddenly Keisha said, “I have an idea. What if we made lots of measuring tools like ours and hooked them together. Then measuring the garden hose would be much easier.”

“I think that will work,” said Rachel. “But if we make the long tool out of paper, it may tear. What else could we use?”

“Let’s go into the basement and see what we can find to make the long measuring tool,” said Grandpa.

In the basement, they found an old window shade that rolled up and down. This is what they did. First, they rolled the window shade out on the floor. Then, they matched the small measuring tool on the shade and traced it with a marker. They matched it end to end until they reached the end of the shade, tracing it each time. Next, they drew the black lines on the measuring tool. They colored each strip a different color. Then, they did this two more times so that they had three long strips, each made up of 20 colored strips. Finally, they cut the long strips out and taped the ends together to make one very long measuring tool.

"Now," said Grandpa, "you have a very long measuring tool. You can measure lots of long things and you won't get tired!" Rachel and Keisha rolled the measuring tool up so it looked like a doughnut and went to measure the garden hose. They measured the hose and found that it was 55 strips long. "Now we have a long strip," said Keisha. "But I don't think it will be long enough to measure the next long object I've thought of."

"Keisha," said Rachel, "what longer thing do you want to measure now?" Keisha laughed and said, "Let's find out how many strips it would take to go around the world!" "Let's do that tomorrow," said Rachel. "I'm too tired today!"

Blackline Master

Measuring Strips

LESSON 16

Using a Measuring Tape to Measure Distance

Overview and Objectives

As they measure distance in this final lesson, students apply many of the skills they have developed during the last 15 lessons, such as using common starting points, using beginning and ending points, and counting by tens. Returning to the activity from Lesson 6 involving the Flippers™, students use their measuring tapes to measure how far they can make the Flippers™ travel. Students expand their awareness that measuring is an extension of matching when they compare the methods used and results recorded in Lesson 6 with those in this lesson. The lesson also serves as an embedded assessment that allows you an opportunity to observe your students as they implement the skills they have learned during the unit.

- Students predict how far they can make the Flippers™ travel.
- Students use their measuring tapes from Lesson 15 to determine the longest distance that the Flippers™ travel.
- Students record their results.
- Students compare the results obtained by measuring with the results obtained by matching.

Background

In this culminating activity, students have an opportunity to apply what they have learned about measuring to measure how far they can make a Flipper™ travel.

During the final discussion, students compare the results they obtain in this lesson with those results obtained in Lesson 6. This challenges students to compare the processes of matching and measuring and to summarize the differences between the two processes.

Materials

For each student

1 measuring tape (from Lesson 15)
1 Flipper™
1 **Record Sheet 16-A: Lengths of Flips**
1 pencil

For the class

1 sheet of newsprint
Student flipping charts (from Lesson 6)
Markers

Preparation

1. Arrange the materials for this lesson in the distribution center.
2. Label the newsprint "Longest Flips with the Flippers." Write each student's name on the chart. Leave space to record their results (see Figure 16-1).

Figure 16-1

Sample chart

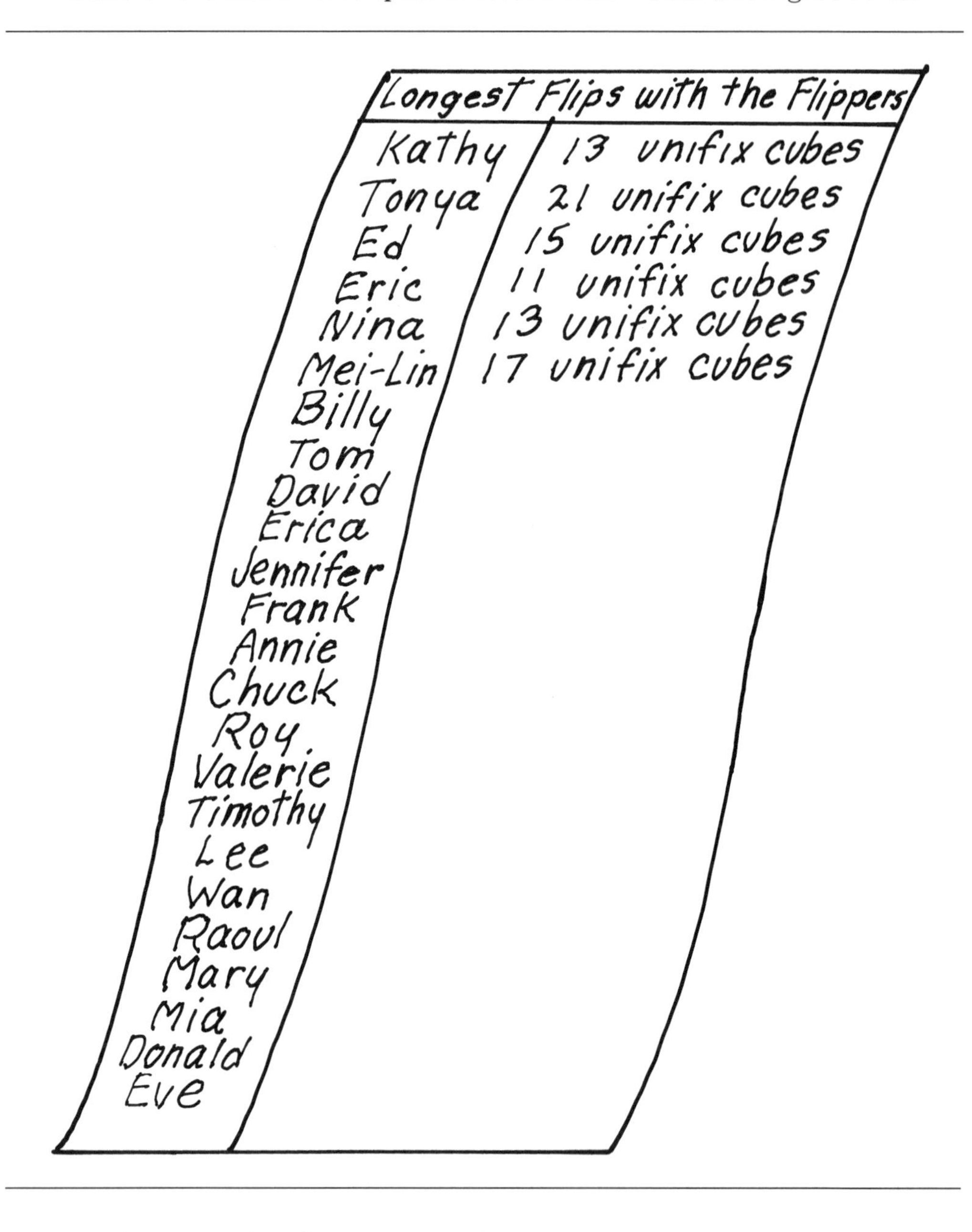

3. Display the student flipping charts from Lesson 6.
4. Copy **Record Sheet 16-A: Lengths of Flips** for each student.
5. Arrange students in pairs. Try to group students with the same partners from Lesson 6.

Procedure

1. Share with students that in today's lesson, they will revisit the Flippers™ activity from Lesson 6. Students will be using their measuring tapes from Lesson 15 to measure the distance of their flips.
2. Have students take a few minutes to discuss with their partners how they used the Flippers™ in Lesson 6. Let students share their thoughts with the class.
3. Pass out and review **Record Sheet 16-A: Lengths of Flips.**

   Note: Have your students decide what they would like to use as a starting line. Some students have used masking tape, a shoe, or a line in the linoleum floor.
4. Let students know the process they will follow.
   - With your partner, collect the materials from the distribution center.
   - Decide what you will use for a starting line.
   - Using your starting line, both you and your partner take turns practicing with your Flipper™. Take five or six practice flips.
5. Now that students have had time practicing with the Flipper™, ask them to do the following:
   - Make a prediction about the distance your Flipper™ will go from the starting line to the spot where it lands.
   - Write your prediction on the record sheet.
   - Now flip your Flipper™. Use your measuring tape to measure the distance from the starting line to the landing point.
   - Record the length of the flip on the record sheet.
   - Switch places and let your partner make a flip with his or her Flipper™.
   - Repeat this activity three more times.
   - Circle the longest flip on your record sheet.
6. Have students collect materials and return them to the distribution center.

Final Activities

1. Have students share the distance of their longest flip with the class. Next to their names, record their results on the chart labeled "Longest Flips with the Flippers."
2. Ask students to focus on the chart and to compare the lengths of the flips. Whose flips were longest? Whose flips were the same? As you made more flips, how did your predictions compare with your actual results?
3. Let students know that they will work together to write a class story about their experiences matching and measuring the distance of the flips.
4. Have students take a few minutes with their partners to compare how they have used the adding machine tape to match the distance of their flips in Lesson 6 and the measuring strip to determine the distance of their flips in this lesson. To help guide the discussion, you may want to use questions such as the following:
   - What types of things did we do to find out the length of objects before we began using the measuring strip?
   - How have our measuring strategies changed since the beginning of the unit?

- What did we do to determine the length of the flips in the earlier lesson (Lesson 6)?
- What did we do in this lesson that was different from the earlier lesson (Lesson 6)?
- What did we find out in this lesson that we did not know before?
- How did the measuring strip change the way we measured?

5. Record their ideas on the sheet of newsprint, forming their thoughts into a story. Try to include a thought from every student.
6. You can use the story several ways. One idea is to make a class "Big Book." Write one sentence of the story on each page and let students illustrate the story. Another idea is to make a small book for each student in your class. Copy the story onto the pages and make one book for each student in your class. Leave enough space on each page of the book so that students can illustrate the story. Have students read their book to another class.

Extensions

MATHEMATICS

1. To extend the idea of measuring distance, students can be "flippers" and measure the lengths of their jumps. They can try different ways of jumping, such as a standing jump or a running jump.

MATHEMATICS LANGUAGE ARTS

2. Students can use their measuring tapes to determine the distance from their classroom to places in the school such as the lunchroom, the media center, or the office. They can then write a book called "Measuring Our School."

MATHEMATICS LANGUAGE ARTS

3. Students can create "My Home" books, including such measurements as the distance from the bedroom to the kitchen.

Assessment

Lesson 16 is an embedded assessment that challenges students to apply much of what they have learned during the unit. To assess students' progress, observe the following areas during the discussion in the **Final Activities:**

- Are students able to identify the strategies they used to measure?
- Do students recognize the changes in their measuring strategies?
- Can students effectively apply the measuring strategies?
- As students gain more experience, are their predictions closer to their actual results?

Post-Unit Assessment

The post-unit assessment (pgs. 129–31) is a matched follow-up to the pre-unit assessment in Lesson 1. Comparing students' pre- and post-unit responses to the same set of questions allows you to document their learning.

Additional Assessments

Additional assessments for this unit are on pgs. 133–35.

Record Sheet 16–A

Name: ______________________

Date: ______________________

Lengths of Flips

1.	2.
________ cubes prediction ________ cubes measurement	________ cubes prediction ________ cubes measurement
3. ________ cubes prediction ________ cubes measurement	4. ________ cubes prediction ________ cubes measurement

Post-Unit Assessment

Overview

This post-unit assessment is matched to the pre-unit assessment in Lesson 1. By comparing the individual and class responses from these activities with those from Lesson 1, you will be able to document and assess students' learning over the course of the unit. During the first lesson, students drew themselves and a partner and wrote about the ways they were alike and different. They also developed two class lists entitled "What We Know about Comparing and Measuring" and "Ways We Are Alike and Different." When they revisit these activities during the post-unit assessment, students are likely to appreciate how much they have learned about comparing and measuring.

Materials

For each student

1 copy of **Record Sheet 1-A: Looking at My Partner and Me** (on pg. 21)
1 package of crayons, including one red crayon and one blue crayon

For every two students

1 resealable plastic bag for collecting materials, 23 × 30 cm (9 × 12 in)

For the class

2 sheets of newsprint
Class lists from Lesson 1: "What We Know about Comparing and Measuring" and "Ways We Are Alike and Different"
1,500 Unifix Cubes™, separated by color
1 container of each of the following:
- 100 wood coffee stirrers
- 100 unsharpened pencils
- 100 plastic spoons
- 100 toothpicks
- 100 small wood spools, 4 cm (1½ in)

15 rolls of adding machine tape
Crayons

Preparation

1. Label one sheet of newsprint "What We Know about Comparing and Measuring" and label the other "Ways We Are Alike and Different." Date the sheets and post them in the classroom.
2. Set up the distribution center in your classroom as you did in Lesson 1.
3. Pair students with the same partners they had in Lesson 1.
4. Copy **Record Sheet 1-A: Looking at My Partner and Me** for each student.

Procedure

1. Ask students to think about what they know about comparing and measuring. After a few minutes, have them share their thoughts with the class. Record these thoughts on the "What We Know about Comparing and Measuring" chart. To help stimulate student discussion, you may want to ask questions such as the following:
   - When have you compared before? When have you measured before?
   - How did you compare? How did you measure?
   - Why were you comparing? Why were you measuring?
2. Let students know you would like partners to decide on one way they are like each other and one way they are different from each other. Invite students to use any materials in the classroom or distribution center to help them find out about their partners.
3. After a few moments, ask students to share their thoughts. To encourage discussion, ask the class questions such as the following:
   - What way are you and your partner alike?
   - How are you and your partner different?
   - Did you use any materials from the distribution center to help make your comparisons?
   - How did these materials help you make comparisons?
4. Record students' thoughts on the "Ways We Are Alike and Different" chart.
5. Ask students to share how they obtained their information. Then display the original lists from Lesson 1. Here are some ways to use the lists to assess student progress:
   - Ask students to identify statements they now know to be true. What experiences during the unit helped them confirm these statements? Asking questions such as "How do you know that?" and "What happened next?" may be helpful.
   - Ask students to correct or improve statements and give reasons for their corrections.
   - Ask students to point out information on their new lists that is not on the original ones.
6. Pass out and review **Record Sheet 1-A: Looking at My Partner and Me.** Then ask students to do the following:
   - Write your name and today's date.
   - Draw a picture of yourself and your partner. Write your partner's name in the box with his or her picture.
   - Draw a red circle around the part of the picture that shows one way you and your partner are alike.

- Draw a blue circle around the part of the picture that shows one way you and your partner are different.
- Write one or two sentences describing each likeness and difference.

7. On the chalkboard, you may want to write sentence starters such as the following:
   - I am like my partner because ________________________.
   - One way I am different from my partner is ________________________.
   - My partner and I ________________________.
8. Invite students to share their drawings with the class.
9. Collect the record sheets and have the students return their materials to the distribution center.
10. As you compare the class lists and record sheets from the post-unit assessment with those from Lesson 1, note the following:
    - Do students' post-unit observations show greater detail than those from Lesson 1?
    - Do students' post-unit comparisons about likenesses and differences include measuring? For example, do students use Unifix Cubes™ to find out the length of their partner's arms?
    - Do students choose standard units of measure? If so, which units do they choose?
    - When students measure, do they use beginning and ending points and a common starting line? Do they label the units in their measurements?

Additional Assessments

Overview

This section presents three suggested assessment activities. Although not essential, they can provide additional information that will help you evaluate student learning. Consider using various kinds of assessments so that students with different learning styles will have opportunities to express their knowledge and skills.

- **Assessment 1** asks individual students to listen to an open-ended measuring problem and to discuss and demonstrate the strategies they would use to solve the problem.
- **Assessment 2** asks students to help create a class story about what they did in the unit.
- **Assessment 3** asks students to use the different tools from this unit to measure their body cutouts from Lesson 2.

Assessment 1: Solving a Measuring Problem

Work with one student at a time for this assessment. As students work on a measuring problem, you can assess whether they have chosen measuring tools appropriate to the task and whether they have used these tools effectively. Remember that in this assessment activity, the emphasis is on **how** students determine the amount of paper to use, not the actual wrapping of the present.

Materials

20 blue Unifix Cubes™
20 red Unifix Cubes™
1 container of each of the following:
- 20 wood coffee stirrers
- 20 plastic spoons
- 20 unsharpened pencils
- 20 small wood spools, 4 cm (1½ in)
- 20 toothpicks

1 measuring strip (from Lesson 13)
1 measuring tape (from Lesson 15)
1 roll of adding machine tape
1 roll of gift wrap or craft paper
1 cardboard shoebox
Scissors
Tape

Note: You can recycle the above materials for each individual assessment.

Procedure

1. Let the student know that in this activity, there will be an opportunity to demonstrate knowledge of measuring to solve a problem.
2. Read the following question: If you had to wrap a shoebox containing a birthday present and wanted to make certain that you had enough paper, what would you do?
3. First, ask the student how to find the answer to the problem. Then show the student the materials that might help solve the problem. Help the student cut the paper and wrap the present, if necessary.
4. Try questions such as the following:
   - What tool did you use to measure? Why did you pick this tool?
   - What did you learn in this unit that helped you select the tool?
   - What did you need to do to make certain that the paper fit the present?
5. If the paper does not fit the box, ask the students what they could do next time to make the paper fit.

Assessment 2: Class Story

Students work as a group to create the class story "What I Did in *Comparing and Measuring.*" Then they complete individual statements about one thing they learned in the unit.

Materials

For the class

Newsprint
Markers
Copy paper

Procedure

1. Let students know that in this activity, they will work together to write a class story.
2. Have students take a few minutes to think about some of the things they did in this unit.
3. Have students share their thoughts with the class. As you record the thoughts on the newsprint, try to form complete sentences without changing the essence of the sentences.
4. You may want to ask questions such as the following:
   - What tools did you use in the unit?
   - How did you use them?
   - What did you find out about your bodies?
   - What did you find out about comparing, matching, and measuring?
5. Record the class story on a blackline master so that it resembles a book. Copy one book for each student.
6. Have students complete the story by adding a sentence about something they learned from the unit.

7. When reviewing the class story and individual student books, you may want to think about the following guidelines:
   - Does the story contain references to using beginning and ending points?
   - Does the story contain references to using a common starting line?
   - Do students recognize the need for standard units of measure?
   - Do the students discuss predictions?

Assessment 3: Measuring Body Cutouts

Work with children individually for this assessment. Each student measures his or her body cutout from Lesson 2. Students use the measuring strips or the measuring tapes made in previous lessons, as well as the other materials listed.

Materials

1 body cutout from Lesson 2
20 blue Unifix Cubes™
20 red Unifix Cubes™
1 container of each of the following:
- 20 wood coffee stirrers
- 20 small wood spools, 4 cm (1½ in)
- 20 unsharpened pencils
- 20 plastic spoons
- 20 toothpicks

1 measuring strip (from Lesson 13)
1 measuring tape (from Lesson 15)
1 roll of adding machine tape

Note: You can recycle the above materials for each individual assessment.

Procedure

1. Review with the student the different measuring units available. Then give the student his or her body cutout from Lesson 2.

   Note: Have a measuring strip and measuring tape on hand in case students no longer have them.

2. Ask the student to select three or four body parts on the cutout to measure. Have the student select the measuring materials he or she would like to use.
3. As each student measures, record your observations about the process the student has followed. Look for evidence of each student's knowledge of measuring by noting the following:
   - Does the student use beginning and ending points?
   - Does the student use iteration?
   - Does the student's reported measurement reflect the actual measurement?
   - Does the student use all the same units to measure a body part or does the student combine units?

Bibliography: Resources for Teachers and Books for Students

This Bibliography provides a sampling of books that complement the unit. It is divided into the following categories:

- Resources for Teachers
- Books for Students

These materials come well recommended. They have been favorably reviewed, and teachers have found them useful.

If a book goes out of print or if you seek additional titles, you may wish to consult the following resources.

Appraisal: Science Books for Young People (The Children's Science Book Review Committee, Boston).

Published quarterly, this periodical reviews new science books available for young people. Each book is reviewed by a librarian and by a scientist. The Children's Science Book Review Committee is sponsored by the Science Education Department of Boston University's School of Education and the New England Roundtable of Children's Librarians.

Gath, Tracy, and Maria Sosa, eds. *Science Books & Films' Best Books for Children, 1992–1995.* Washington, DC: American Association for the Advancement of Science, 1996.

This volume, part of a continuing series, is a compilation of the most highly rated science books that have been reviewed recently in the periodical *Science Books & Films.*

National Science Resources Center. *Resources for Teaching Elementary School Science.* Washington, DC: National Academy Press, 1996.

This guide, a completely revised edition of *Science for Children: Resources for Teachers,* provides extensive information about some 350 hands-on, inquiry-centered science curriculum materials for grades K–6. It also annotates other published materials—books on teaching science, science book lists, and periodicals for teachers and students. The guide includes annotated listings of museums and federal and professional organizations throughout the country with programs and other resources to assist in the teaching of elementary school science.

Science and Children (National Science Teachers Association, Arlington, VA).

Each March, this monthly periodical provides an annotated bibliography of outstanding science trade books primarily for elementary students.

Science Books and Films (American Association for the Advancement of Science, Washington, DC).

Published nine times a year, this periodical offers critical reviews of a wide range of new science materials, from books to audiovisual materials to electronic resources. The reviews are primarily written by scientists and science educators. *Science Books and Films* is useful for librarians, media specialists, curriculum supervisors, science teachers, and others responsible for recommending and purchasing scientific materials.

Scientific American (Scientific American, Inc., New York).

Each December in this monthly periodical, Philip and Phylis Morrison compile and review a selection of outstanding new science books for children.

Resources for Teachers

Charlesworth, Rosalind, and Karen K. Lind. *Math and Science for Young Children*. New York: Delmar Publishers, Inc., 1990.

This well-written book is designed to be used by teachers wishing to teach math and science on the basis of a developmental sequence of learning. Activities are suggested for presenting numerous math and science skills, topics, and concepts.

Dishon, Dee, and Wilson O'Leary. *A Guidebook for Cooperative Learning: Techniques for Creating More Effective Schools.* Holmes Beach, FL: Learning Publications, 1984.

A practical guide to help teachers implement cooperative learning in the classroom.

Johnson, David W., Roger T. Johnson, and Edythe Johnson Holubec. *Circles of Learning: Cooperation in the Classroom.* Alexandria, VA: Association for Supervision and Curriculum Development, 1984.

Presents the case for cooperative learning in a concise and readable form. Reviews the research, outlines implementation strategies, and answers many questions.

Labinowicz, Ed. *The Piaget Primer: Thinking-Learning-Teaching*. Illustrated by Susie Pollard Frazee. Menlo Park, CA: Addison-Wesley, 1980.

Suitable for both the novice and expert teacher, this is a detailed but readable text about child growth and development. The author uses practical examples to illustrate his ideas.

Books for Students

Ferber, Elizabeth. *Once I Was Very Small.* Toronto, Canada: Annick Press, 1993.

An easy-to-read story about growing bigger.

Irving, Nicole. *The Biggest.* London: Usborne Publishing, 1987.

Tom and Sally go looking for some of the biggest animals and things in the world.

Lionni, Leo. *Inch by Inch.* New York: Astor-Honor, 1960.

An inchworm keeps from being eaten by a robin by proving he is useful for measuring things.

Most, Bernard. *How Big Were the Dinosaurs?* San Diego, CA: Harcourt Brace, 1994.

Gives children a relative idea of the size of different kinds of dinosaurs by comparing them with objects in the modern world.

Myller, Rolf. *How Big Is a Foot?* New York: Dell, 1990.

A simple, amusing story illustrating the importance of using standard units of measure.

Schwartz, David M. *How Much Is a Million?* New York: Scholastic, 1985.

Marvelosissimo the Mathematical Magician finds unusual ways to quantify a million, a billion, and a trillion.

Spier, Peter. *People.* New York: Doubleday, 1988.

Explores the many ways people are different from one another.

Wylie, Joanne, and David Wylie. *A Big Fish Story.* Chicago: Children's Press, 1983.

An easy-to-read picture book to help children learn about size.

National Science Resources Center Advisory Board